Earth on Fire

Earth on Fire

How volcanoes shape our planet

By Bernhard Edmaier
Text by Angelika Jung-Hüttl

Gorely, Kamchatka Peninsula, Russia
There are many active volcanoes on Russia's Kamchatka Peninsula. The funnel-shaped crater of the Gorely volcano is visible in the foreground.

Contents

Introduction

Angelika Jung-Hüttl

Every time a volcano erupts, we are violently reminded that we live on a fireball. Only a thin crust of rock separates us from the furnace deep inside our planet. At its core, at a depth of 6,370 kilometres, it is as hot as the surface of the sun: 5,000 degrees Celsius. At 1,000 kilometres deep, the temperature is around 2,000 degrees Celsius, and about 100 kilometres below the earth's surface, it is still between 800 and 1,200 degrees Celsius. At these temperatures, stone and rock melt.

Up to a quarter of the heat inside the earth originates from the time our planet came into existence, over 4.6 billion years ago, when a cloud of cosmic dust condensed to form the primordial earth, which was gradually heated up by meteorite collisions and the bombardment of stray chunks of rock from space. Three quarters of the heat is created through the continual decay of radioactive elements.

Despite the high temperatures, the rock under the earth's surface is not all as fluid as the magma that rises through a volcano's vent and gushes out of its crater as lava. On the contrary: the centre of the earth's core is solid, as is around 99 per cent of the mantle of the earth. Nevertheless, the hot material of the mantle is not hard and brittle like the rock on the earth's surface, but flexible. It circulates slowly in gigantic convection streams following the laws of thermodynamics, constantly moving from hot to cold areas. The reason for the plastic nature of this rocky material is predominantly the enormous pressure, which averages around 350,000 bar at a depth of 1,000 kilometres and climbs to around 4 million bar at the earth's centre. However, the different melting points of the various minerals in the interior of our planet also play a role.

How magma, liquid molten rock, ultimately develops in the earth's interior has not yet been fully scientifically explained, but hot gases, water and shifts in pressure are important factors in this complex process. Researchers have more exact ideas about where magma is found in the earth's interior. The analyses of seismic waves that run through the globe following earthquakes as well as nuclear explosions offer key insights, as these waves travel more slowly in fluid or partly melted rock than they do in solid rock. Not only the earth's crust consists of solid rock, but also the upper layer of the earth's mantle. Scientists call this hard layer of rock – which is only a few kilometres thick below the oceans, but up to 200 kilometres thick underneath the continents – the lithosphere, which means brittle or rock shell. Under this layer the rock is flexible and is called the asthenosphere, or soft shell.

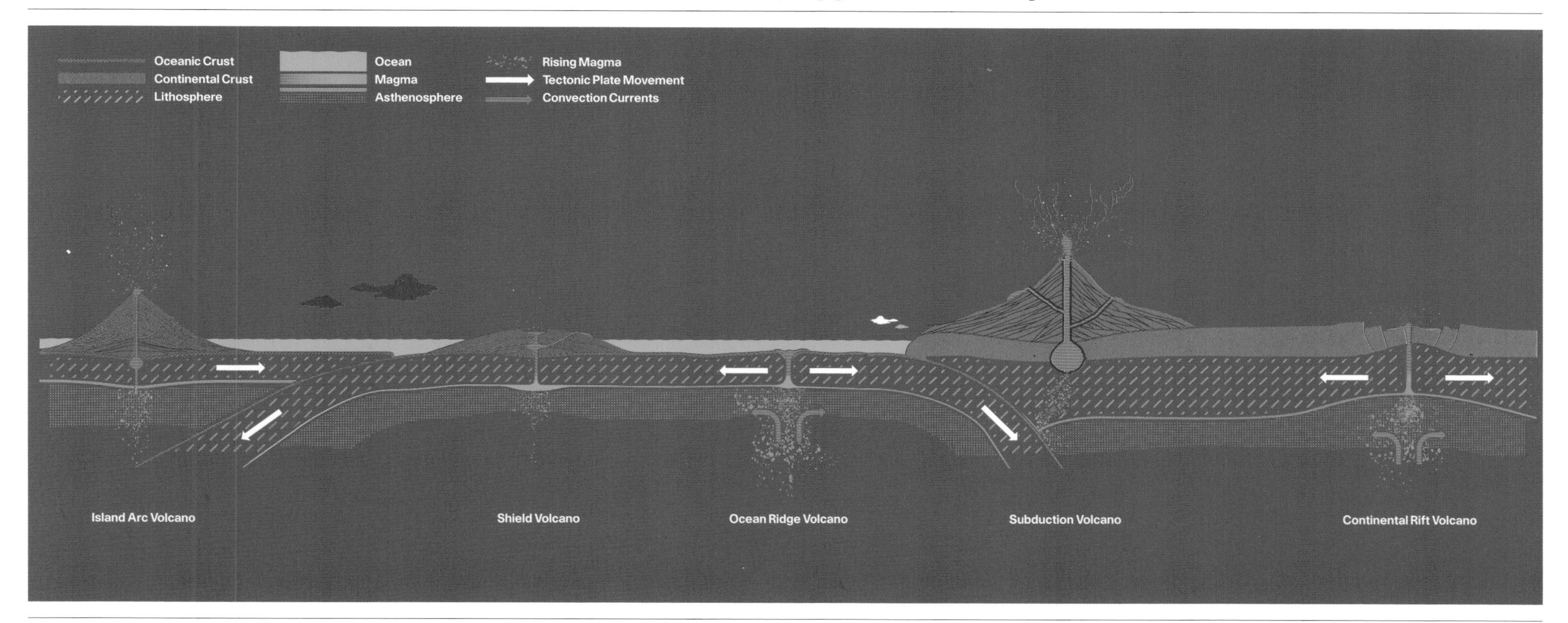

Tungurahua, Ecuador
The steep cone of the Tungurahua volcano in Ecuador is 5,023 metres high (opposite). During its most recent larger eruption, in February 2008, many of the fields on its western flank were destroyed by heavy ashfall. The families living there were safely evacuated.

Types of volcano
The heat and pressure in the earth's interior make the rock in the asthenosphere soft and malleable. On top of the asthenosphere lies the lithosphere, our planet's hard shell of brittle rock. This shell has broken into a number of plates that float on the asthenosphere, colliding against and diving underneath each other. Glowing molten rock or magma can only break through the hard, rocky crust of the earth at certain points: it mainly streams out along mid-ocean ridges through submarine fissures, pushing the lithosphere apart along a wide front, thereby forming ocean ridge volcanoes. Along subduction zones, where plates which have been forced downwards start to melt, rising magma produces island-arc volcanoes, such as those of the Indonesian island chain from Sumatera to Flores, or subduction volcanoes such as those of the North American Cascades or the South American Andes, like Tungurahua in Ecuador (opposite). Hot upsurges of magma feed hot spot or shield volcanoes such as the volcanoes of Hawaii. Rising magma can also cause contintents to break apart, leading to continental rift volcanoes such as those of the East African Rift Valley.

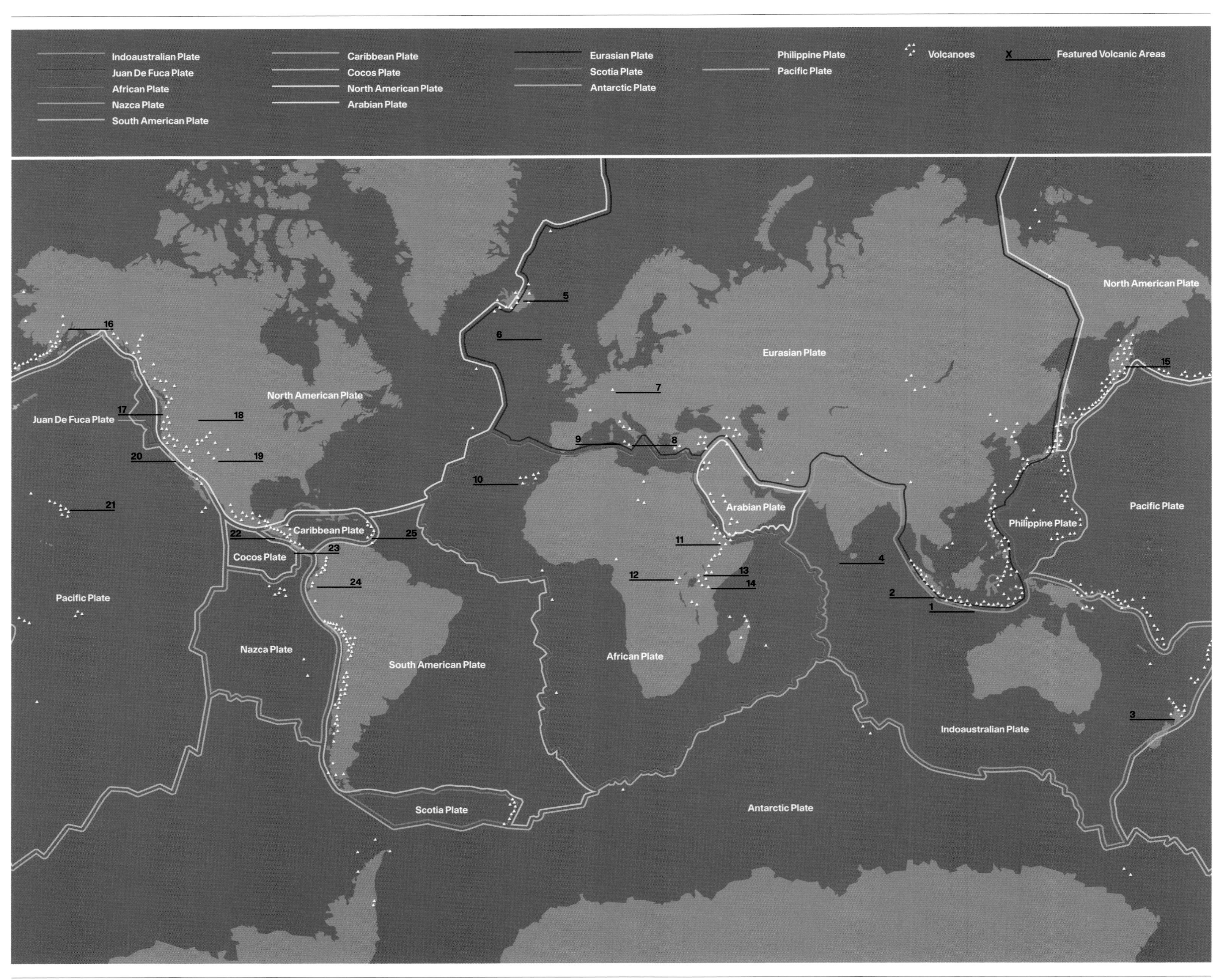

Featured volcanoes and locations

1 **Bali, Indonesia**
Gunung Agung, 10
Batur, 26

2 **Java, Indonesia**
Krakatau, 13, 204
Kawah Ijen, 14
Merapi, 16, 17
Dieng Plateau, 23
Bledug Kuwu, 25

3 **North Island, New Zealand**
Tongariro, 15
Tikitere, 18
Champagne Pool, 19–21
Wai-O-Tapu, 22, 24
White Island, 27–8
Waimangu Valley, 208
Whakarewarewa, 208, 210
Ruapehu, 210

4 **Maldives atolls**
29–31, 212–3

5 **Iceland**
Kjos valley, 32
Thingvellir, 55
Landeyjarsandur, 57
Mývatn 58
Bjarnarey Island, 59
Hekla, 60
Reykjanesviti, 61
Laki fissure, 62
Krafla, 63, 68
Hveravellir geothermal area, 65
Svartsengi, Reykjanes Peninsula, 66–7
Veidivötn, 69
Brennisteinsalda, 71
Snæfellsness, 72
Mællifellsandur, 73
Svartifoss, 74
Skeidararjökull, 75
Námafjall geothermal area, 77
Skeidararsandur, 199

6 **Faroe Islands, Denmark**
54

7 **Germany**
Pulvermaar, Eifel, 193

8 **Italy**
Etna, 34–9, 41–3, 45–9
Vesuvius, 192

9 **Aeolian Islands, Italy**
Stromboli, 50, 193, 198
Vulcano, 51

10 **Canary Islands, Spain**
Montañas del Fuego, Lanzarote, 52, 53
La Geria, Lanzarote, 198

11 **Djibouti and Ethiopia**
Assal Rift, 78
Erta Ale, 81, 93, 99, 105, 107
Awash River Flood Plain, 87
Afar depression, 88
Lac Assal, 91
Dallol, 92, 100, 112–3
Dala Fila, 96
Ghoubbet al Kharab, 97
Sawabi Islands 102
Lac Abbé, 106

12 **Democratic Republic of Congo**
Nyamulagira, 83–5, 115

13 **Kenya**
Marsabit, 80, 94–5
Nabuyaton crater, 89
Lake Turkana, 103
Sugata Valley, 108–9
Lake Magadi, 110

14 **Tanzania**
Ol Doinyo Lengai, 86, 101
Lake Natron, 111

15 **Kamchatka Peninsula, Russia**
Gorely, 4, 191
Mutnovsky, 118–9, 211
Avachinsky, 120
Koryaksky, 120
Maly Semiachik, 121
Karymsky, 190
Kliuchevskoi, 205

16 **Alaska, USA**
Iljamna, 132
Mount Wrangell, 137
Mount Spurr, 139

17 **Washington and Oregon, USA**
Mount St Helens, 122–3, 128, 206
John Day area, 145–7
Crater Lake caldera, 150–1
Newberry volcano, 152, 191

18 **Wyoming, USA**
Yellowstone caldera, 124, 127, 129–31, 144, 149, 153

19 **New Mexico, USA**
Ship Rock, 125

20 **California, USA**
Ubehebe crater, 140
Death Valley, 141

21 **Hawaii, USA**
Kilauea, 116, 133, 135–6, 142–3, 188–9
Na Pali Coast, Kawaii, 155

22 **Guatemala**
Pacaya, 166–7, 169, 171
Santa María, 195
Santiaguito, 177
Fuego, 178, 204

23 **Costa Rica**
Poás, 156, 164–5
Botos, 174–5
Arenal, 160, 194
Rincón de la Vieja, 161–3, 170

24 **Ecuador**
Tungurahua, 6, 159
Cotopaxi, 172
Guagua Pichincha, 173

25 **Montserrat**
Soufrière Hills, 179, 181–3, 196

Tectonic plates
The earth's hard rocky shell, the lithosphere, resembles a gigantic jigsaw puzzle. It is broken into many pieces, the tectonic plates, which drift slowly on the planet's hot, molten interior. Magma can rise up at the joints between these plates, and this is where most volcanoes are located. However, there are also volcanoes in the middle of plates; 'hot spots' fed by magma springs, known as mantle plumes, in the depths of the earth.

The perfect pressure and temperature conditions for rock to melt are found at the intersection between the lithosphere and asthenosphere, at a depth of around 75 to 250 kilometres. According to the latest findings, this is where most of the magma that is pushed up to the earth's surface and builds up in magma chambers underneath volcanoes is produced.

The fluid molten mass cannot, however, break through the hard rocky shell of the earth at any arbitrary point. It uses the weak areas in the lithosphere, which, according to tectonic theory, is divided into seven large and 13 small plates. The individual plates sit together tightly, like a jigsaw puzzle, and are constantly on the move, carried along on the convection currents in the asthenosphere as if on a conveyer belt. As a result of this motion they are stretched and compressed, rub up against each other, collide, dive under or even move apart.

Magma primarily pushes up to the earth's surface at the points where lithospheric plates adjoin. Along these seams and fissures in our planet's surface, volcanoes are arranged one beside another like pearls on a necklace. Only a few volcanoes lie on the plates themselves, and their sources of magma generally lie in the very depths of the earth. Scientists believe that this molten rock's point of origin lies 2,900 kilometres under the earth's surface, in the border area between the earth's mantle and its core. They refer to a hot upsurge, or mantle plume, in the earth's interior that melts through the lithosphere from below like a blowtorch, creating a 'hot spot' on the earth's surface.

Volcanoes are the valves of the earth, opening when the pressure of the magma pushing up to the earth's surface becomes too strong. Lava-spitting mountains are not the only evidence of heat and powerful movement in the depths of the earth: there are also geothermal areas, with their hot springs and bubbling mud pools, steaming acid lakes in often perfectly circular craters, fountains of corrosive gases, as well as deposits of bright-yellow sulphur crystals and multi-coloured soils that owe their many hues to the presence of a wide variety of minerals.

Huge cracks and fissures in the earth, many kilometres long, are also the result of volcanic activity, as are caves and subterranean tunnels within the earth's crust. The black sand on the beaches of volcanic islands is nothing more than volcanic rock ground down by the waves, while coral atolls grow on the cones of once-active volcanoes that have been worn down by the sea and gradually disappeared underneath the surface of the ocean.

Glowing lava flows out not only on land, but also under the sea – in fact far more so than out of the craters on the continents. Eighty per cent of all lava emerges from the fissures and craters along the plate boundaries that cross the ocean floors. Over millions of years, these flows have built the vast mountain ranges of the mid-ocean ridges that rise 3,000 metres above the ocean floor, are 1,000 kilometres wide and span the entire globe, with a total length of 65,000 kilometres. They run from north to south through the middle of the Atlantic, stretching out across the floor of the Indian Ocean to the Red Sea, dividing into multiple branches in the Eastern Pacific. When underwater volcanoes break the surface of the sea, volcanic islands can be created, like Hawaii in the Pacific, or Iceland, the Azores, St Helena and Tristan da Cunha in the Atlantic.

Of all the active continental volcanoes, around 20 are spitting out lava and ash at any one time, but eruptions usually only make headlines if human lives are endangered. More than 500 million people, around a tenth of the world's population, live in areas affected by active volcanoes. Since 1700, the date from which we have detailed records of catastrophic volcanic eruptions, around 260,000 have lost their lives to them.

The great ancient philosophers already sought scientific explanations for these at once fascinating and devastating forces of nature. Plato, for instance, explained volcanic eruptions as battles between the three elements of air, fire and earth. When air escaped from the fire inside the earth, he surmised, the earth shook and volcanoes broke out. Air also played an important role in Aristotle's theory. He believed that surging coastal waves pushed air into caves deep inside the earth, where it was set alight by tar and sulphur. According to the philosopher, flames and smoke gushed out of the vents in the volcanoes as a result, and ash shot out of their craters.

However, most people at the time held giants, demons, angry gods or the devil responsible for volcanic eruptions. Sacrifices were offered to appease them, sometimes even human sacrifices. Even today, elements of these superstitions survive in many countries. Hawaiians, for example, lay fruit and flowers in front of glowing lava flows to placate the fire goddess Pele. In Sicily, the residents of the communities near Mount Etna organize processions whenever the volcano erupts, and beg their patron saints to keep their houses and fields safe.

Gunung Agung, Bali, Indonesia
The giant funnel of Gunung Agung, one of the most dangerous volcanoes in Indonesia, measures 500 metres in diameter. The huge bowl of the Batur caldera opens up in the background.

The Indoaustralian Plate is one of seven large continental plates that, along with many smaller fragments, make up the hard stone shell of the earth. Almost three quarters of it are covered by oceans. Apart from some groups of islands, only two large landmasses protrude above the water surface – Australia and the Indian subcontinent.

This giant plate drifts at a speed of around seven centimetres a year towards the northeast, where it collides with its large neighbours, the Eurasian and Pacific Plates. The best-known results of this collision are the Himalayas, which were upfolded as India bored into Eurasia. At the Malay Archipelago, the Indoaustralian Plate is pushed into the earth's hot interior and melts. Magma resurfaces in a chain of more than one hundred, mostly highly explosive, volcanoes that stretches from Sumatra to the Lesser Sunda Islands.

In the southeast, the Pacific Plate pushes under the Indoaustralian Plate. New Zealand's volcanoes are among those that mark this seam, where the underground heat feeds a range of hot springs as well as one of the largest geyser fields in the world.

The Maldives, a chain of islands at the southern tip of India, also belong to the Indoaustralian Plate. Over 1,000 small coral islands and atolls rest on an ancient volcanic mountain range that sank after the volcanoes became extinct.

Indoaustralia

Anak Krakatau
Sunda Strait, Indonesia

The nine-kilometre long, five-kilometre wide and 1,000-metre high volcanic island of Krakatau once stood where Anak Krakatau now propels ash into the air. On 27 August 1883, after an eruption lasting two days, Krakatau disappeared in the waters of the Sunda Strait between Sumatra and Java. Eruption clouds shot 50 kilometres into the atmosphere, blocking out the sun, and tidal waves killed 36,000 people in the first volcanic disaster to be reported worldwide via the then new telegraph system. Forty years later, fishermen spotted gas bubbles in the sea where Krakatau had vanished. On 12 August 1930, eruption clouds first pierced the surface of the water and Anak Krakatau, the child of the sunken volcano, was born. Today, it is around 320 metres high – and still growing.

Kawah Ijen
East Java, Indonesia

The fumaroles that jet out of the crater wall of Kawah Ijen reach temperatures of up to 580 degrees Celsius. Pure sulphur is deposited around the exit holes in quantities large enough to support mining. Clouds of white steam hide workers chipping away lumps of the bright yellow mineral, with damp cloths they wrap around their mouths and noses as their only protection from the corrosive gases.

Sulphur gas has turned the water in the crater of the volcano, which has a temperature of around 40 degrees Celsius, into sulphuric acid. Gas explosions can occur in the green, corrosive lake, posing danger not only to the miners but also to the farmers in the coffee plantations on the volcano's flanks, should this water spill over. A dam was built at the crater wall's lowest point to protect them.

Mount Tongariro
North Island, New Zealand

Tongariro is not a single volcano, but a whole volcanic complex. Different eruption phases have formed several overlapping craters. The most attractive is Red Crater, whose blood-red walls indicate that the rock there is rich in iron that has oxidized at high temperatures. Its flanks are adorned by the Emerald Lakes, two explosion craters filled with rainwater.

About three kilometres away, the perfectly symmetrical cone of Mount Ngauruhoe rises up, the highest peak in the Tongariro massif at 2,291 metres. It is only 2,500 years old. Mount Ruapehu stands in the distance, covered in ice. At 2,797 metres, it is the tallest mountain on New Zealand's North Island and one of the country's most active volcanoes, notorious for its mudslides.

Mount Merapi
Central Java, Indonesia

At least three of the many volcanoes in Indonesia are called Merapi, meaning 'mountain of fire'. The best-known is the 2,968-metre high Merapi in the centre of the island of Java, one of the most dangerous of the nation's 129 volcanoes. More than one million people live in the immediate area, and the city of Yogyakarta, Indonesia's cultural capital, is only 30 kilometres away.

The gas cloud that constantly pours from the summit is clearly visible in the cool evening air. Around 3,000 tonnes of carbon dioxide, 400 tonnes of sulphur dioxide, 250 tonnes of hydrochloric acid and 50 tonnes of hydrofluoric acid are expelled daily. If the composition of these gases changes, scientists are on high alert, as it could indicate an impending eruption.

Mount Merapi
Central Java, Indonesia

What makes this stratovolcano so dangerous is the plug of hot rock slowly pushing out of its vent, creating a lava dome. Pieces of rock break off from it and crash down the flanks of the volcano. By day, these falling rocks can usually only be heard, but at dusk the glowing pieces of lava that bounce down the slope and explode become visible. Whole families travel to the observatory in Babadan, just five kilometres from the summit as the crow flies, to witness the fireworks.

However, every few years the lava dome collapses. People in the immediate danger zone are evacuated as clouds of ash shoot out of the crater and pyroclastic flows charge down the slopes. During the rainy season these fresh emissions can turn into potentially dangerous lahars and mudslides.

Hell's Gate
Tikitere geothermal region, North Island, New Zealand

Bubbly foam floats on top of the hot grey water in the mud pools of Tikitere, and new bubbles are constantly emerging from its invisible depths. When they burst, it smells of rotten eggs – the stench of hydrogen sulphide. This poisonous, corrosive gas, which escapes from the magma deep under this geothermal area, stings your nose and makes your eyes water. It's best to keep your distance from these bubbling pools.

This area is sacred to the Maori, the indigenous people of New Zealand. It is named after a princess who was married to a terrible chieftain against her will. Fleeing from him, she threw herself into one of the hot corrosive pools and died.

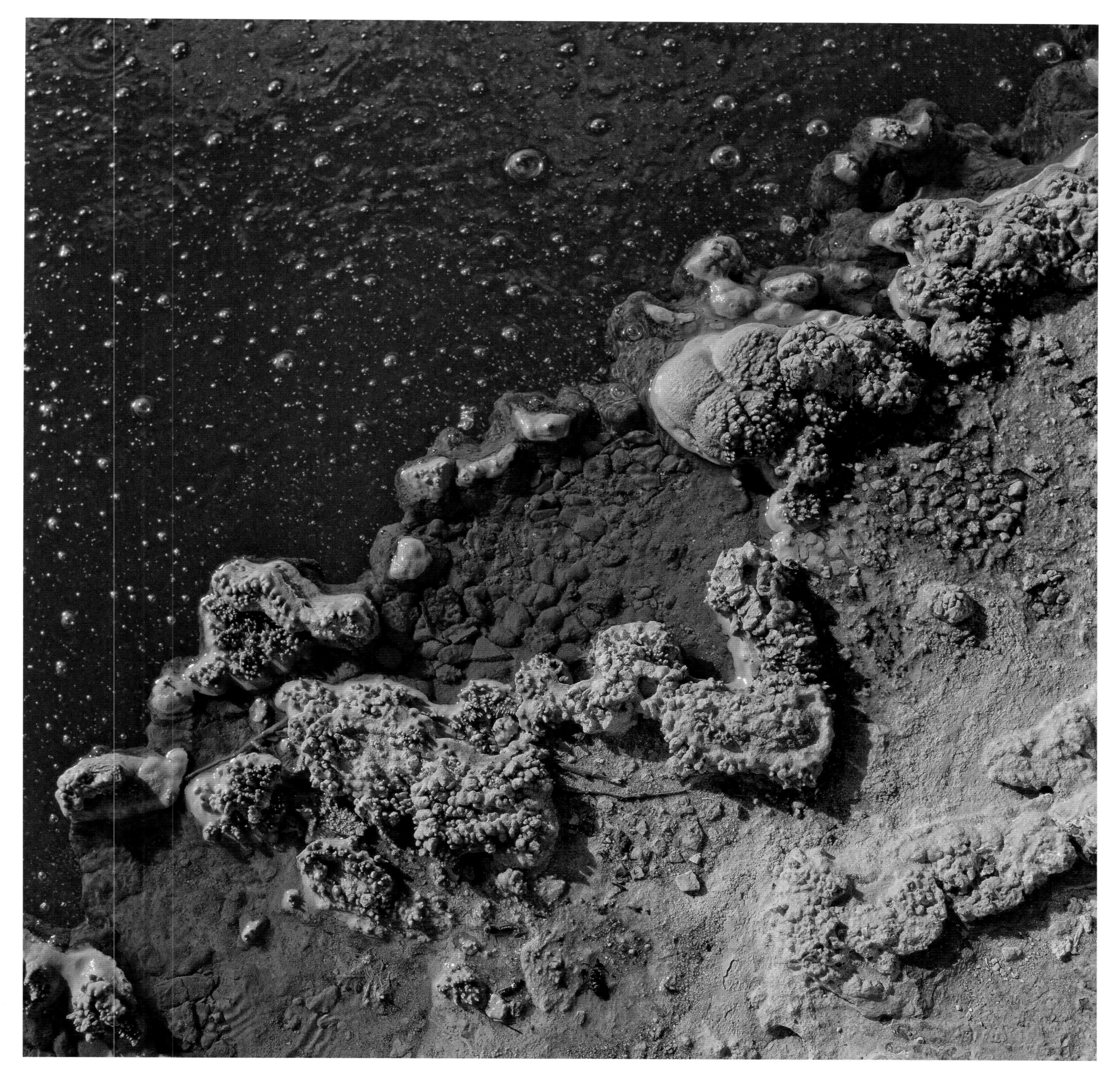

Champagne Pool
Wai-O-Tapu geothermal region, North Island, New Zealand

A thick sinter crust surrounds the Champagne Pool, a 60-metre deep crater torn open during a volcanic explosion in Wai-O-Tapu 900 years ago. The spring water in the basin is heated to 75 degrees Celsius, and heat-loving bacteria cover the crust, where minerals are deposited. The orange colour is evidence of antimony compounds, but traces of mercury, silver and even gold can also be found. Hot water from the pool flows over the Artist's Palette, a brittle platform coloured yellow by iron minerals and sulphur (see also following pages).

Bubbles of carbon dioxide fizz to the surface of the green, chlorine-rich mineral water and burst, giving the pool its name. The gas and steam create billowing mists that gradually dissipate into the air.

Sulphur crystals
Wai-O-Tapu geothermal region, North Island, New Zealand

Thick, bright-yellow crusts of sulphur often form where fumaroles gush from holes and fissures in volcanic rock, as soon as the hot vapours meet the cold air. At temperatures of over 95 degrees Celsius, sulphur crystallizes into fine needles, creating very fragile forms that can be shattered by a light breeze. At 110 degrees Celsius, the sulphur liquefies and trickles out of fissures in the volcanic rock in bright-yellow, orange or sometimes even brown streams, depending on the temperature.

Exploding mud bubble, Sikidang
Dieng Plateau, Central Java, Indonesia

The residents of the Dieng Plateau call this crater filled with mud that spatters in all directions Sikidang, which means 'jumping deer'. As if someone had lit a fire beneath the crater, the dark brew bubbles and churns, catapulting droplets of mud many metres into the air. The viscous gloop is hot but not at boiling point, so this is not caused by heat from below but by gas bubbles that shoot up from the floor of the crater and burst on reaching the surface of the pool.

There are also fumaroles and acid lakes on the Dieng Plateau. This volcanic complex, consisting of a number of craters like Sikidang, is notorious for its phreatic eruptions, in which hot mud and steam are suddenly thrown into the air in fountains many metres high.

Bubbling mud
Wai-O-Tapu geothermal region, North Island, New Zealand

The drier the season, the more viscous the brew in the mud pools of Wai-O-Tapu. Every bursting gas bubble leaves behind a concentric circular bulge in the mud, as well as a small cloud of stinking sulphuric gas.

Wai-O-Tapu – the name comes from the Maori language and means 'holy water' – is the most spectacular of the 129 geothermal fields on New Zealand's North Island, where temperatures at a depth of five kilometres can be as high as 350 degrees Celsius. That is more than twice as hot as in non-volcanic regions, where the ground heats up by three degrees for every 100 metres one drills down vertically. Volcanic heat brings the ground water in the earth up to the boil and creates hot springs and bubbling mud pools on the surface.

Bledug Kuwu mud geyser
Central Java, Indonesia

The bubbles of the Bledug Kuwu geyser grow up to 10 metres high before they suddenly explode, sending shreds of mud flying through the air and releasing a white cloud of carbon dioxide. Strictly speaking, Bledug Kuwu is not a geyser, which is by definition powered by overheated ground water and blows out fountains of hot steam. The driving force of this mud geyser is upwelling volcanic carbon dioxide, although its exact mechanism is not yet clear.

The ground near the mud geyser is springy under foot but, if you get too close, you will slowly sink into the lukewarm, grey mass. Apart from carbon dioxide, the geyser also ejects mineral water, particularly in the rainy season, which local villagers boil down to a sweetish salt they can sell at a high price.

Mount Batur
Bali, Indonesia

Mount Batur is a stratovolcano consisting of three interlocking cones. It lies in a caldera about 10 kilometres in diameter that was created during a massive eruption around 25,000 years ago. Its highest crater towers over the floor of the caldera by some 700 metres.

Mount Batur erupts every few years – most recently in 1994, 1998 and 2000, although these eruptions were all moderate. Clouds of ash shot into the air and streams of lava poured down the volcano's flanks. Batur's magma is created at a depth of 150 kilometres beneath the active craters. This is where the Indoaustralian Plate melts, having been pushed under the Eurasian Plate in the region of the Indonesian island arc.

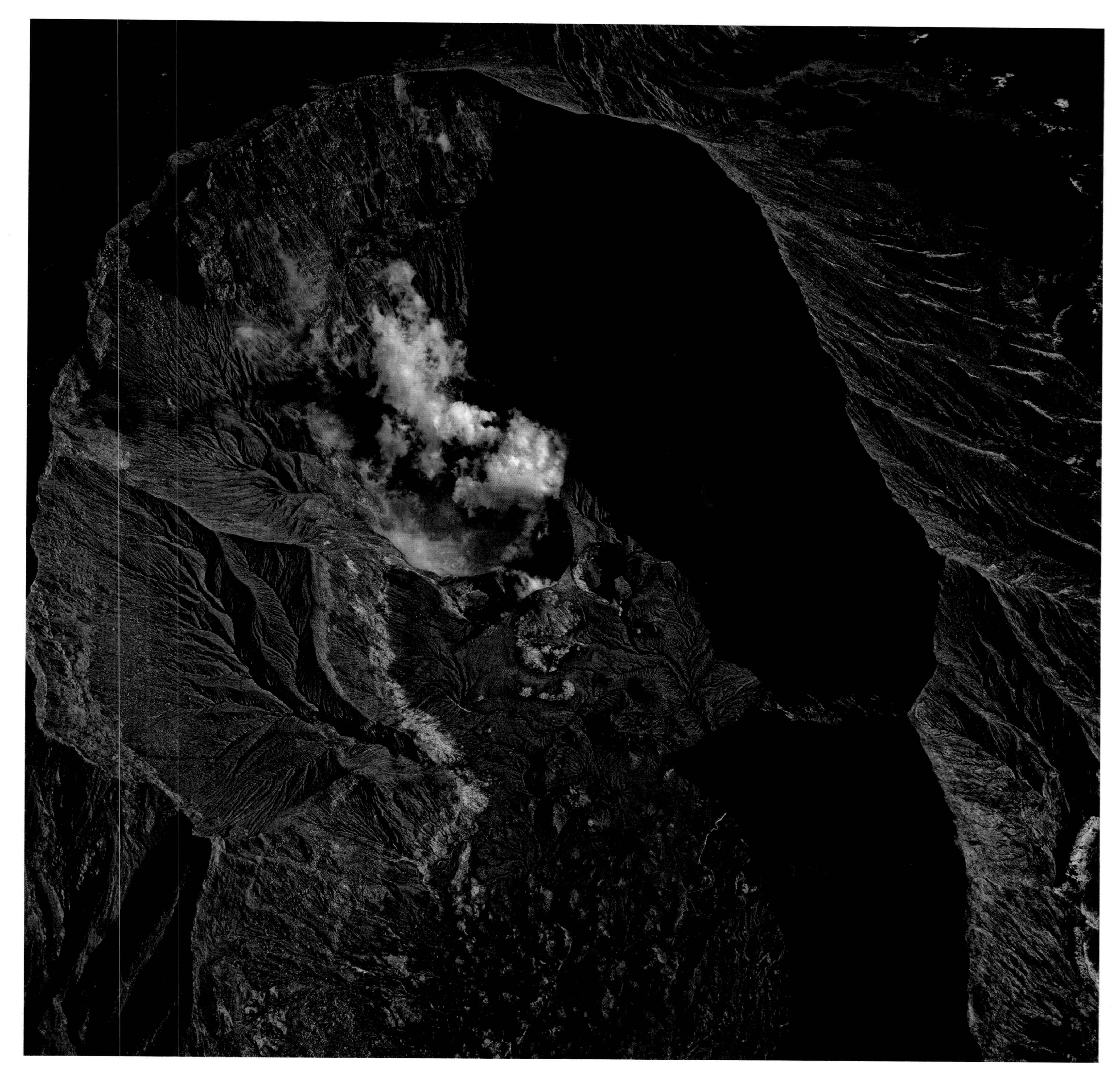

White Island
Bay of Plenty, New Zealand

The crater of the White Island volcano measures between 200 and 250 metres in diameter. An acid lake simmers at its centre, and rainwater has carved out wide branching gullies in the crater walls. Fumaroles gush out of the crater floor, rising up into the air as gas clouds, while yellow sulphur is deposited around the outlet holes.

White Island's horseshoe shape dates back to September 1924, when part of the crater wall caved in during a violent eruption. Sulphur was mined on the island at the time, but the mine was destroyed during the eruption and buried beneath the rubble of the crater wall. All 11 miners lost their lives and, since then, the island has remained uninhabited.

White Island
Bay of Plenty, New Zealand

Like all volcanoes in New Zealand, White Island is a product of subduction. Deep in the earth's interior, the Pacific Plate is pushing under the Indo-australian Plate, where it melts. A lot of water is dragged down with the plate, so eruptions in this region are often highly explosive.

White Island was discovered in 1769 by the British explorer Captain James Cook (1728–79) on his first South Sea voyage. Its volcanic activity has only been documented since 1826, since when there have been 35 eruptions, among them the catastrophic event during which part of the crater wall collapsed. The island lies 48 kilometres off New Zealand's north coast. Measured from the sea-floor, the volcano is around 1,000 metres high, but only 321 metres are above sea level.

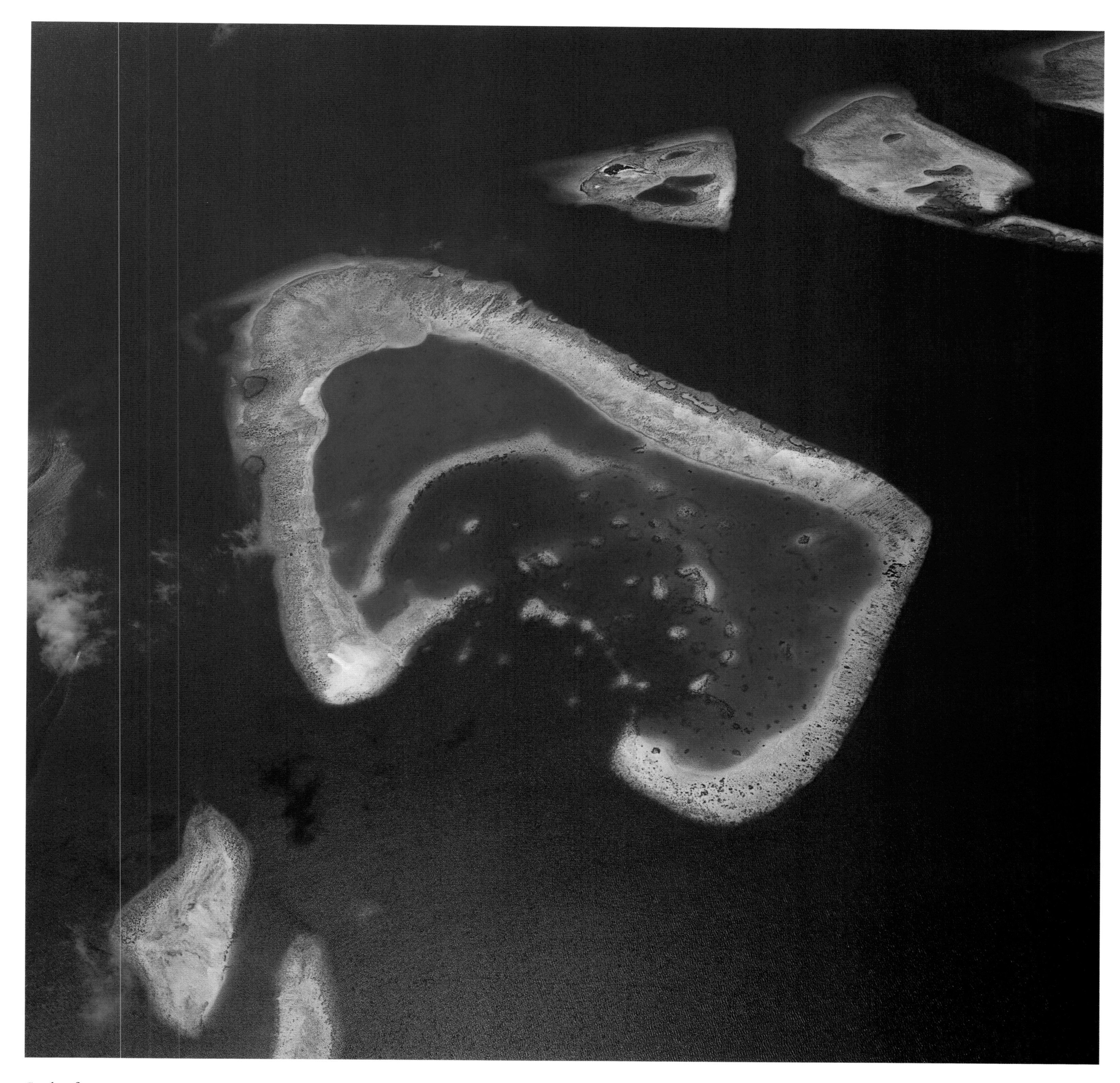

Coral reef
North Male Atoll, Maldives

Volcanic islands in tropical waters are colonized by corals. First a pale ridge appears around the coast – a fringing reef. When the volcano becomes extinct, it is eroded by wind and water and disappears into the sea. The coral, however, keeps growing, so once the volcanic cone has sunk, an atoll is created – a circular reef with a lagoon in the middle. If the wind regularly blows from one direction, surging waves can break through this ring.

The Maldives sit on an old volcanic mountain range that sank into the sea millions of years ago. Not every little reef ring is above a crater – the long-extinct volcanoes are covered by a layer of coral limestone 2,000 metres thick, on which small new reef islands develop, often disappearing again during heavy storms.

South Male Atoll
Maldives

The approximately 1,190 islands of the Maldives are spread over 25 large atolls. South Male Atoll is about 40 kilometres long and 20 kilometres wide. The Maldives were shaped not only by the volcanic mountains below, but also by the ice age. When much of the earth's water was held on land as ice, large atolls rose out of the ocean as limestone platforms, and became weathered. When the sea level rose after the ice age, they were submerged again.

The outer rings of large atolls like South Male are broken by many channels. Fresh, clear water at a temperature of 20 to 30 degrees Celsius flows through these gullies into the lagoons, creating perfect conditions for coral, and many small, often circular, reefs and coral islands have grown here since the ice age.

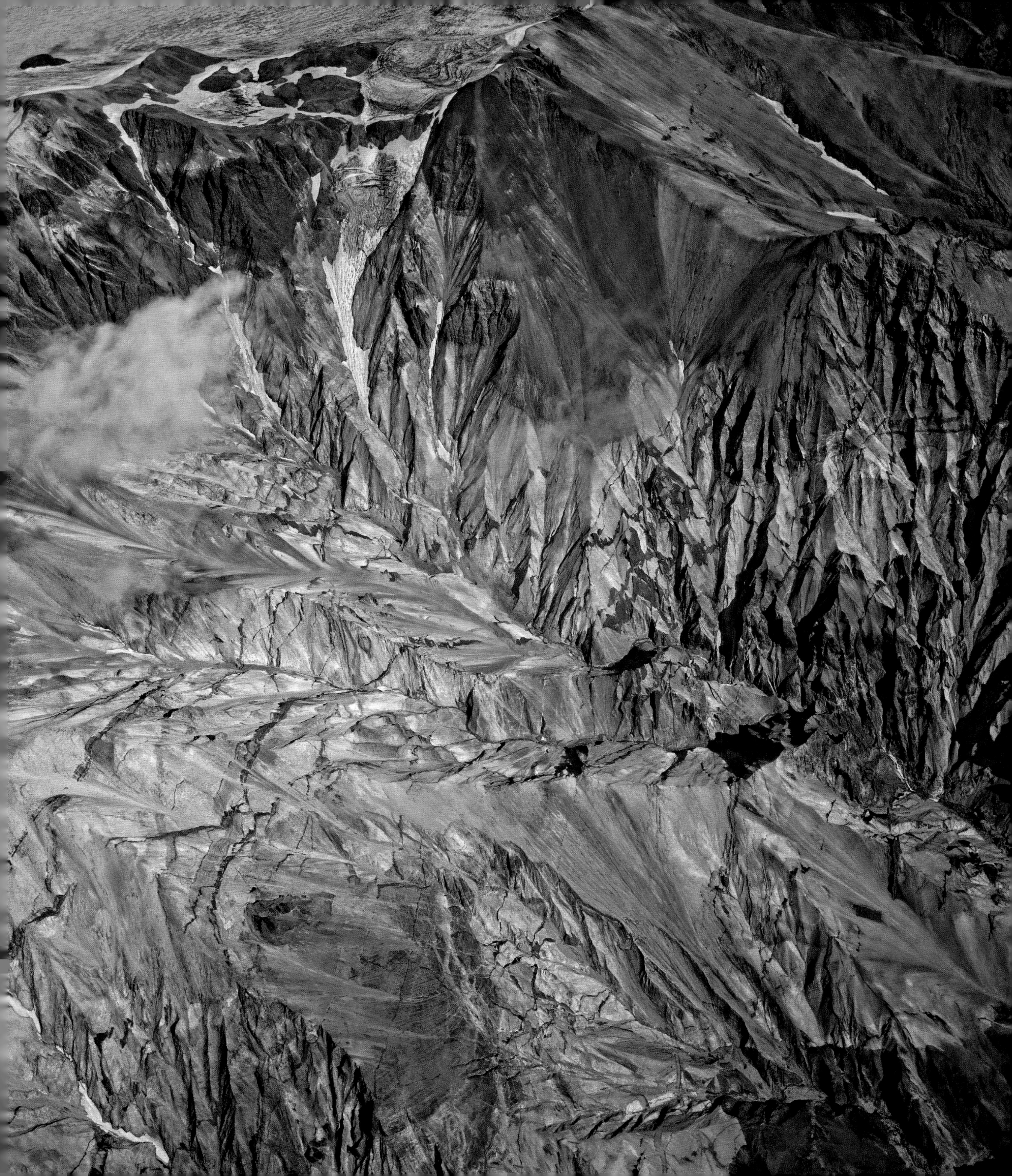

Kjos, Iceland
The Kjos valley lies on the southern edge of Vatnajökull, Europe's largest glacier, whose ice sheet covers a number of active volcanoes. The valley cuts deep into the flanks of a former subglacial volcanic system. The once charcoal-grey stone has been attacked by corrosive, sulphuric gases and turned a browny yellow.

On primordial earth, the eastern border of what we now know as Europe crashed into Asia, a landmass four times its size, and they fused together to make Eurasia, one of the seven large tectonic plates of our modern-day planet. The long, narrow mountain range of the Urals rises up along this ancient, long-since fused boundary, running south from Russia's north coast for over 2,000 kilometres.

Geologically speaking, Europe shares its western border with North America. The two plates meet in the middle of the Atlantic, where the mountains of the Mid-Atlantic Ridge run along the sea-floor from north to south, deep under the ocean. Magma surges up from the earth's interior through fissures in the sea-bed here, and pushes the two giant plates apart. Europe and North America move between five and 10 centimetres further away from each other every year. The only place where this plate boundary is visible above water is in Iceland, where a hot spot presses the Mid-Atlantic Ridge upwards, breaking the surface.

To the south, the African Plate is pushing under Europe. This massive, driving force has left spectacular traces on both sides of the plate boundary. At the northern edge of Africa the Atlas Mountains have been created. The extent of this range is visible to the west in the form of the Canary Islands, their active volcanoes rising above the surface of the Atlantic. At the southern edge of Europe are volcanoes whose eruptions were already being documented in ancient times, among them Mount Etna on Sicily, one of the most active volcanoes in the world.

Europe

Sulphur crust
The Chasm, Mount Etna, Sicily, Italy

Hot corrosive gases stream out of the ground at the edge of Mount Etna's craters. As soon as they come into contact with cool air, a variety of minerals are deposited. Pure sulphur creates patches of colour on the dark solidified lava, ranging from pale yellow to orange. The white crusts consist of gypsum (calcium sulphate), salmiac (ammonium chloride) and even salt (sodium chloride).The ground around these crusts is warm and moist.

If the gas runs out, because the supply either fails or has found another way out through the pores and fissures in the rock, this magnificent display quickly disappears; the fine crystals disintegrate or dissolve.

Mount Etna
Sicily, Italy

Etna has four craters at its summit. Since the eruption of September 1986, the highest point of Mount Etna has been at the edge of the Northeast Crater (in the foreground) at 3,350 metres. The Voragine Grande, known in English as the Chasm, opens up next to it. It has a diameter of around 300 metres, like its neighbouring crater to the southwest, the Bocca Nova. In 1971 a fourth vent opened up at the foot of the other three craters – the Southeast Crater. The dark ash that covers the snow on the flanks of the three large craters like a black veil comes from this vent. This aerial shot was taken in June 2000, since when the Southeast Crater has grown around 150 metres in height, almost equalling the highest point of Mount Etna.

Mount Etna
Sicily, Italy

Ash clouds rise up hundreds of metres from a new crater hole that opened up on the southern flank of Mount Etna at the end of October 2002, only a few days before this picture was taken. At the same time, the volcano expelled clouds of steam from its peak craters. This eruption lasted until March 2003, during which time the Torre del Filosofo mountain hut was buried in ejected material.

At 3,350 metres, Mount Etna is the highest volcano in Europe and one of the most active. The eruption of 2002–3, which lasted only five months, was one of the most violent of the last 350 years.

Mount Etna
Sicily, Italy

For five months during the eruption of 2002–3, Mount Etna spewed vast quantities of ash out of the crater that had opened up on its southern flank, around 500 metres below the summit. The cities and villages in the area surrounding Etna suffered recurring spells of falling ash.

Within hours of the eruption, five centimetres of ash had already fallen on the small town of Nicolosi, around 15 kilometres away. Traffic ground the granules into fine dust and stirred it up into the air. People began to have breathing difficulties, and snow ploughs were used to clear the streets of the dark ash. The airport in Catania, 30 kilometres away, was closed for days.

Mount Etna
Sicily, Italy

Although it only lasted for a few days, the July/August 2001 eruption of Etna was more explosive and threw out more ash than any eruption for hundreds of years. Seven fissures in total opened up on the mountain's northeastern and southern flanks. The crater that opened up on the south side, not far from the mountain station of Sapienza, was particularly spectacular. During the eruption, glowing shreds of lava were catapulted into the air and a mighty stream of viscous lava flowed down into the valley. This eruption destroyed an access road to the mountain station as well as a cable car.

Mount Etna
Sicily, Italy

The activity of the crater that opened up on the southern flank of the volcano at the beginning of Mount Etna's October 2002 eruption was particularly spectacular by night. In the dark, the glowing nuclei of clouds of ash became visible. At the time this photograph was taken in November 2002, the cinder cone was around 50 metres high. Lumps of lava up to 50 centimetres in diameter, also known as bombs, were catapulted out of the vent with the ash, smashing deep pits into the soft volcanic discharge. These bombs can be deadly to anyone too close to an eruption zone.

Mount Etna
Sicily, Italy

The spectacle presented by the Southeast Crater of Etna between 1996 and 1998 was quite moderate by its usual standards. Like a fireworks display, lumps of lava and ash particles shot 30 metres high out of its vent, for a few minutes at a time, over and over again. The spiral flight path of the projectiles can be seen in this shot from February 1998. This type of eruption is known as strombolian, after the Italian volcano Stromboli.

The Southeast Crater formed at the foot of the three older summit craters in 1971. To begin with, it was just a small, black mound, but from 1978 it grew higher during generally very violent eruption phases, often connected to lava eruptions on the volcano's lower flanks. Today it is over 200 metres high.

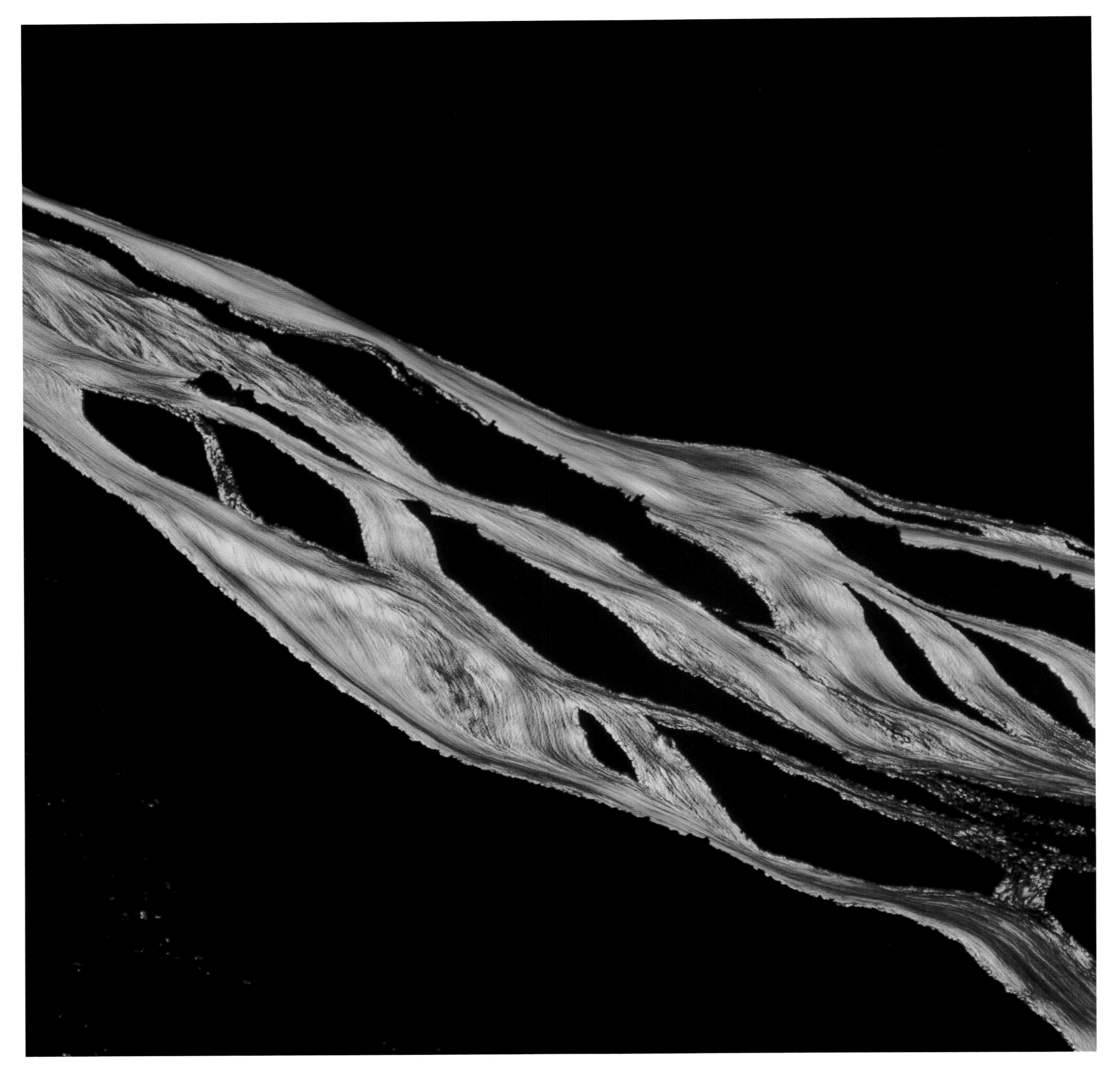

Mount Etna
Sicily, Italy

During the eruption phase that began in December 1991 and lasted until March 1993, Mount Etna did not throw out much ash, but instead ejected vast quantities of lava. This photograph dates from March 1992. The molten rock oozed out of a new fissure in the rock face of the Valle del Bove, a steep, wide cut in the eastern flank of the volcano. On average, five to six cubic metres per second flowed out of the new opening, peaking at around 30 cubic metres per second. The glowing streams of lava, visible from many kilometres away by night, flowed hundreds of metres into the valley. The molten rock came very close to the small town of Zafferana, and a few orchards on the eastern side of the town were buried under the glowing river.

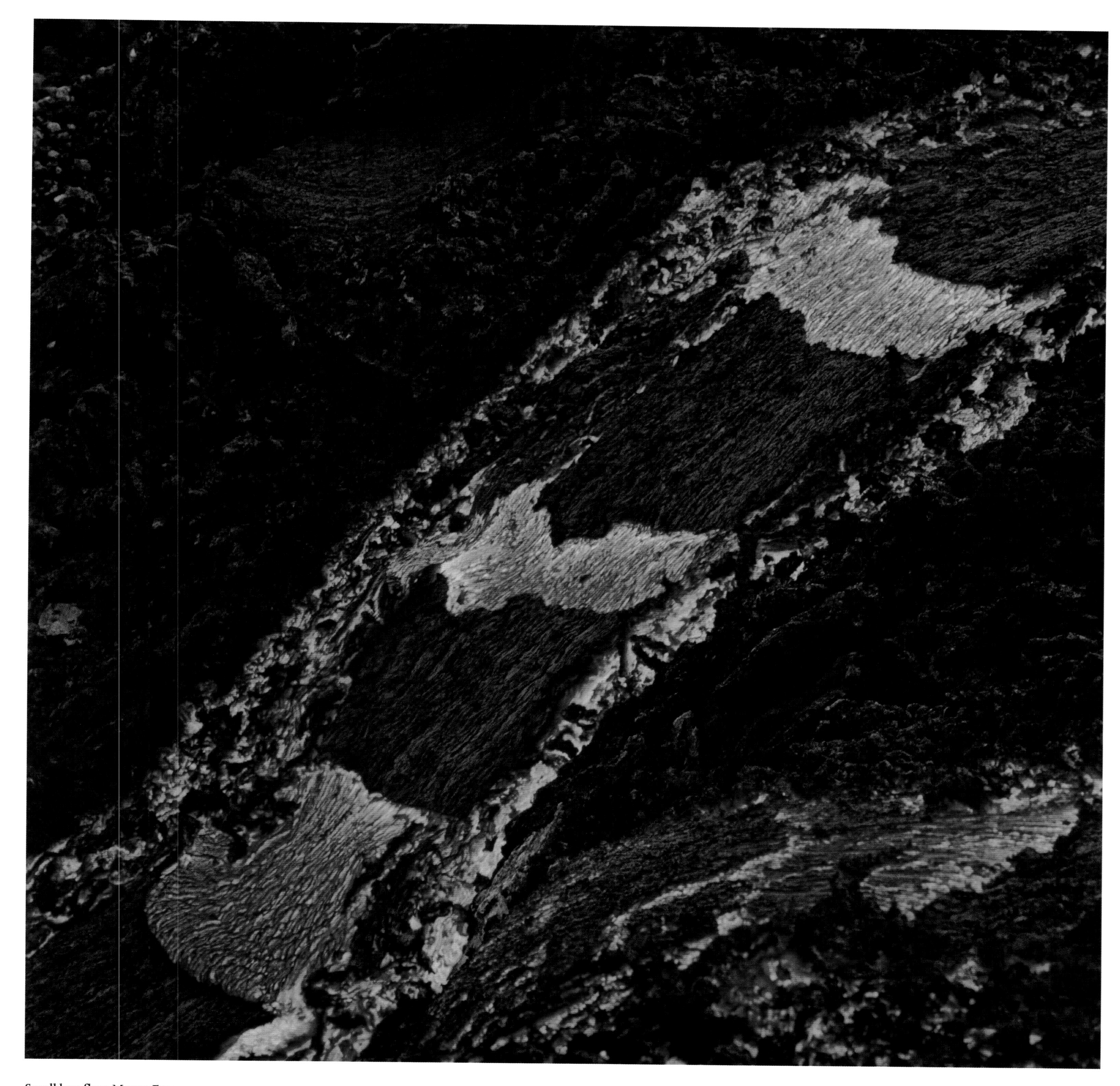

Small lava flow, Mount Etna
Sicily, Italy

The glowing lava of Etna reaches temperatures of 1,000 to 1,100 degrees Celsius. The further the molten rock travels from its exit point, the more it cools down, gradually turning black. This small lava stream had already developed a surface crust when it flowed over steeper terrain, thereby speeding up its flow and ripping open the crust. The glowing stream carries the dark plates down the slope as if on a conveyer belt. The large plates of the lithosphere, the earth's hard shell, drift along on the fluid material of the asthenosphere in a similar way.

Mount Etna
Sicily, Italy

The clouds of ash and gas that climbed into the air from Etna's summit craters and newly-created openings in October 2002 reached a height of just over six kilometres. There they were picked up by the winds, drifting as far as Greece in westerlies, and over Malta to Algeria in Africa in northwesterlies. Satellite pictures showed that these drifting ash clouds were up to 1,300 kilometres long.

The photograph of the crater at the summit of Etna on the following pages was taken in February 1998, four years before the large eruption of 2002. Sulphurous steam is shown pouring out of the Bocca Nova, which is only separated from the Chasm by a narrow ridge. The small Southeast Crater is black from the fresh ash it catapulted out of its mouth at regular intervals.

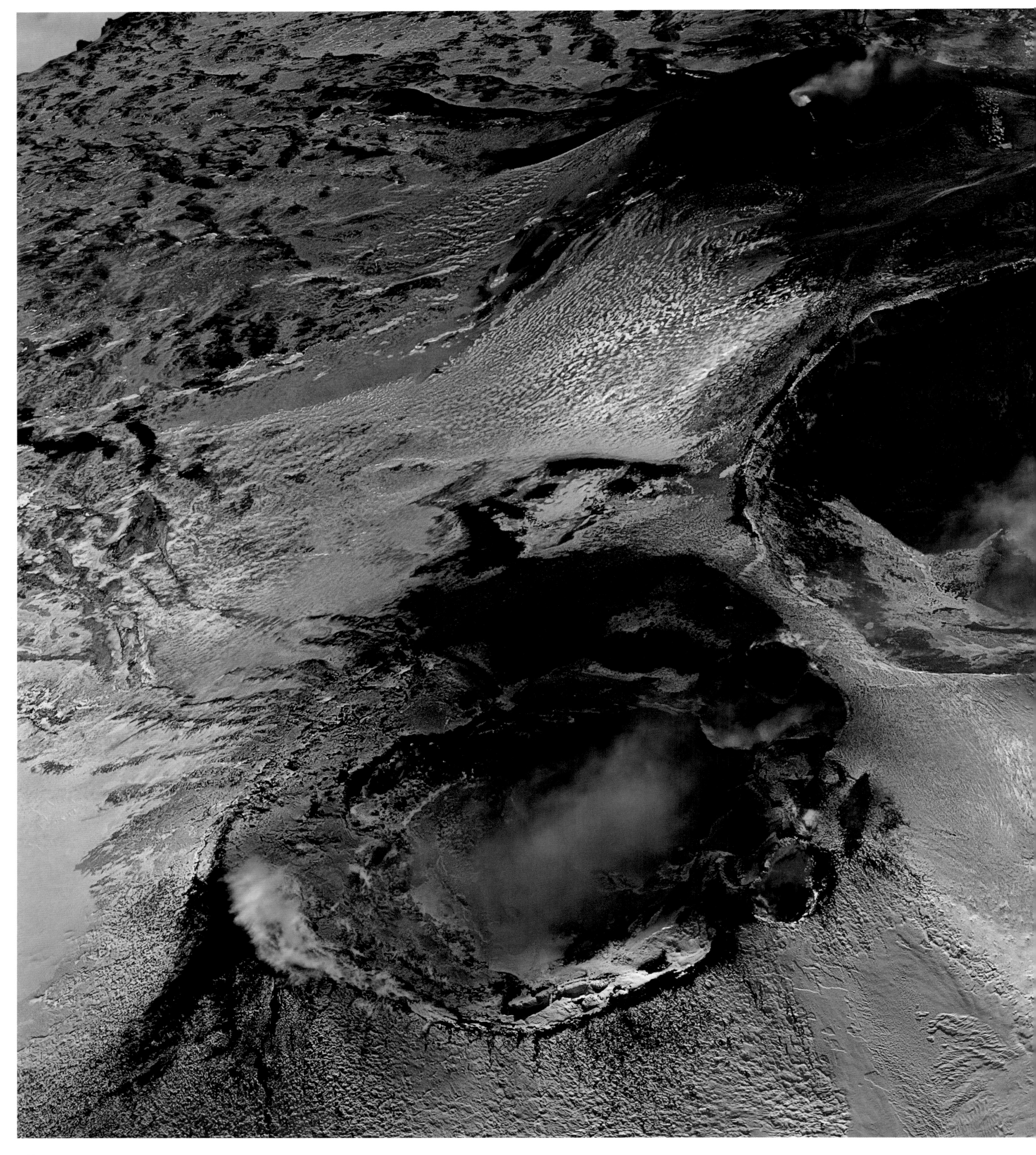

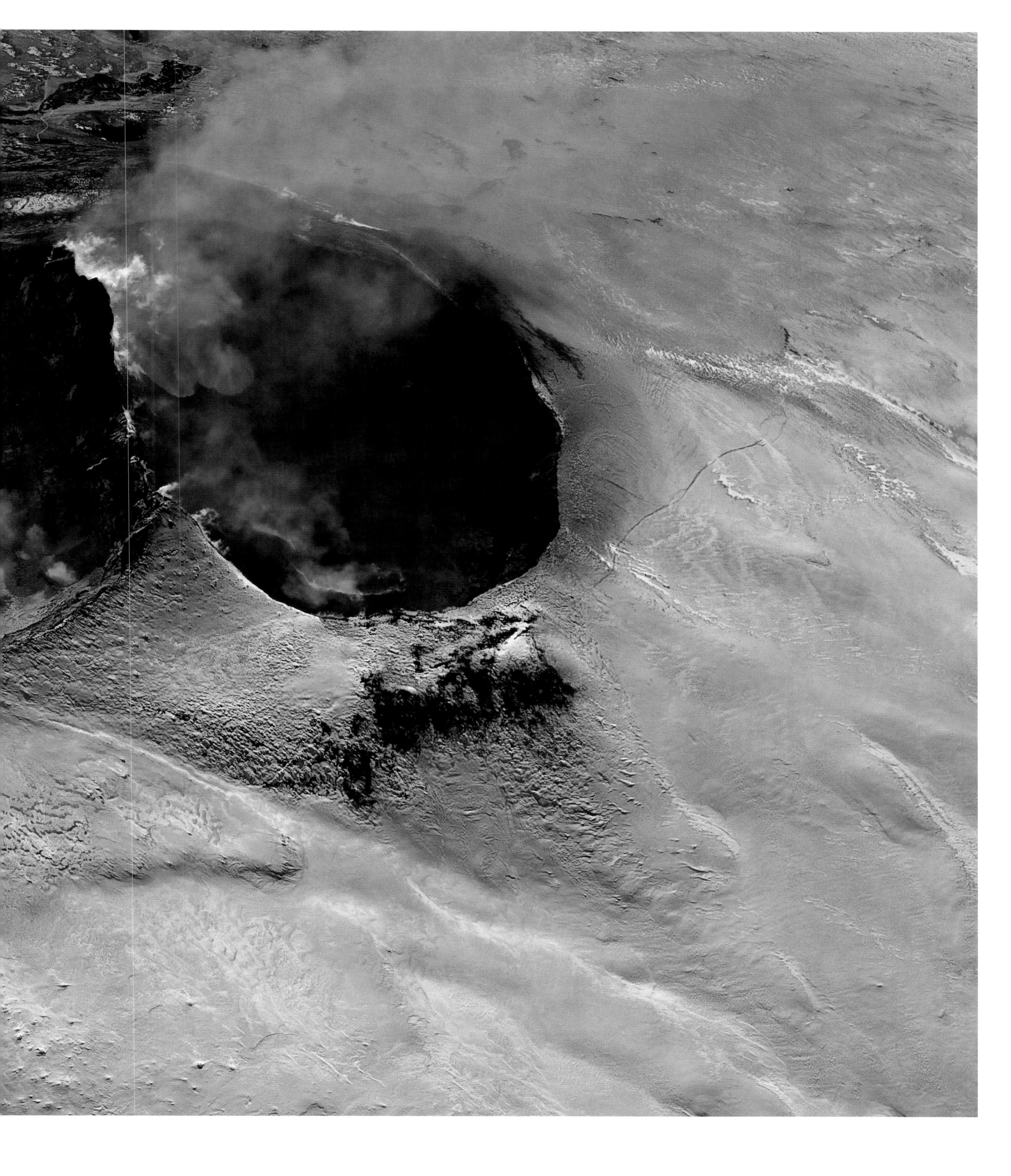

Mount Etna
Sicily, Italy

The slopes of Mount Etna are strewn with cinder cones. More than 270 such hills, some of them up to 250 metres high, cover the volcano, most notably its southern and western slopes. Scientists call these parasitic cones. Single cones are rare; they are usually found in groups, arranged along fissures that have opened up during various flank eruptions.

Despite Mount Etna's location in the south of Italy, its summit is covered in snow from October to June. At its base, it is warm enough all year round for tropical plants to flourish.

Mount Etna
Sicily, Italy

For a few days in September 2004, fumaroles suddenly began to pour out of the lava rock below Etna's steaming Southeast Crater, and left behind a carpet of dark yellow sulphur. At the same time, a small fissure opened up a little further down the slope, out of which a small lava stream flowed. It dried up in March 2005, before the molten rock could reach any populated areas. This photograph was taken in October 2004.

The Saracens who conquered Sicily from Africa in the ninth century AD used the Arabic word for mountain – *jebel* – for Mount Etna. The Sicilians combined this expression with the Latin for mountain – *mons* – and have used the name Mongibello, the 'mountain of mountains', for Etna ever since.

Stromboli
Aeolian Islands, Italy

Stromboli has been active constantly throughout human history, shooting fountains of ash and lumps of lava up to 200 metres into the sky a few times every hour. It belongs to the Aeolian Islands, a group of volcanic islands between Sicily and mainland Italy. Its summit towers 924 metres above sea level, while its currently active crater lies on a plateau some 100 to 150 metres lower down.

Stromboli's coast is populated, and this volcano has hit the headlines in recent years as its eruptions have become more violent. Lava flows kept streaming out of its northwest flank into the sea, and in December 2002, earthquakes set off a landslide. This led to a small tsunami that capsized boats and flooded houses along the coast. The island was evacuated for two months.

Fossa crater
Vulcano, Aeolian Islands, Italy

Fumaroles stream out over the crater edge of the volcano Fossa on Vulcano, a neighbouring island of Stromboli. The gases reach temperatures of 100 to 675 degrees Celsius and deposit sulphur, gypsum, halite (sodium chloride) and alum (hydrated aluminium potassium sulphate), as well as salmiac (ammonium chloride) and realgar, an intensely orange-coloured arsenic sulphide mineral.

The ancient Romans believed this island to be the home of the fire god Vulcan, which is why it and all lava-spewing mountains on earth are now named after him. The Romans mined alum here, using it to tan animal hides. The deposits from Vulcano's hot gases were still being mined even in the late nineteenth century, when work in the corrosive steam was mostly done by convicts.

Montañas del Fuego
Lanzarote, Canary Islands, Spain

Iron minerals give the old volcanic cone Caldera de los Cuervos, or Raven's Crater, its intense colour. The crater is encircled by dark lava flows dating back to the last major eruptions on Lanzarote, between 1730 and 1736 and in 1824. The thin, runny lava that poured out of the ground flooded an area of around 200 square kilometres – a quarter of the whole island.

Lanzarote belongs to the Canary Islands, which lie off the northwest coast of Africa in the Atlantic ocean, around 1,000 kilometres from Spain. All the islands are volcanic in origin, and their volcanoes are fed by a hot spot in the earth's interior.

Montañas del Fuego
Lanzarote, Canary Islands, Spain

'On the 1st of September 1730 the earth broke open between 9 and 10 o'clock at night, and by the next morning a mountain of considerable height had already formed. Just a few days later a second chasm opened up ...'. So begins the eyewitness report of the vicar in the village of Yaiza on Lanzarote, describing the creation of the Montañas del Fuego, the youngest volcanoes on the island. Over time, lava penetrated the fields of the island's inhabitants and eventually reached the sea, killing many fish. When after a few months there still seemed no end to the lava pouring out of an ever-increasing number of new craters, most people left the island. The eruption lasted five years in total. Today, the desert-like region is a national park.

Faroe Islands
North Atlantic, Denmark

Rising 800 metres above the North Atlantic, the strange, steep cliffs of the Faroe Islands, which lie about halfway between the north of Scotland and Iceland, are made up of volcanic rock. Seventy million years ago, fissure eruptions took place here, first underneath the ocean and later above sea level. Vast amounts of lava streamed out, creating a plateau 3,000 to 4,000 metres thick. As the North Atlantic began to open up, this plateau broke and was eroded. The Faroe Islands are all that remains. Today, the only indication of past volcanic activity is a mineral spring on one of the islands, which has a water temperature of 20 degrees Celsius.

Thingvellir
Iceland

The massive trenches crisscrossing Iceland are signs that the earth's crust is being stretched. The fracture zone running across this North Atlantic island divides it into two parts: the eastern section belongs to the European Plate, the western to the North American Plate. Iceland is part of an underwater mountain range, or to be more precise, a section of the Mid-Atlantic Ridge, which is being pushed above sea level by a hot spot – an upflow of magma from inside the earth. Most of the island's more than 30 active volcanoes are not perfect cones, but huge flat shields, or mountain ridges built up on long splits in the earth's crust.

The Almannagjá trench is a historic area: Iceland's old chieftains held their meetings here some 1,000 years ago because of the good acoustics.

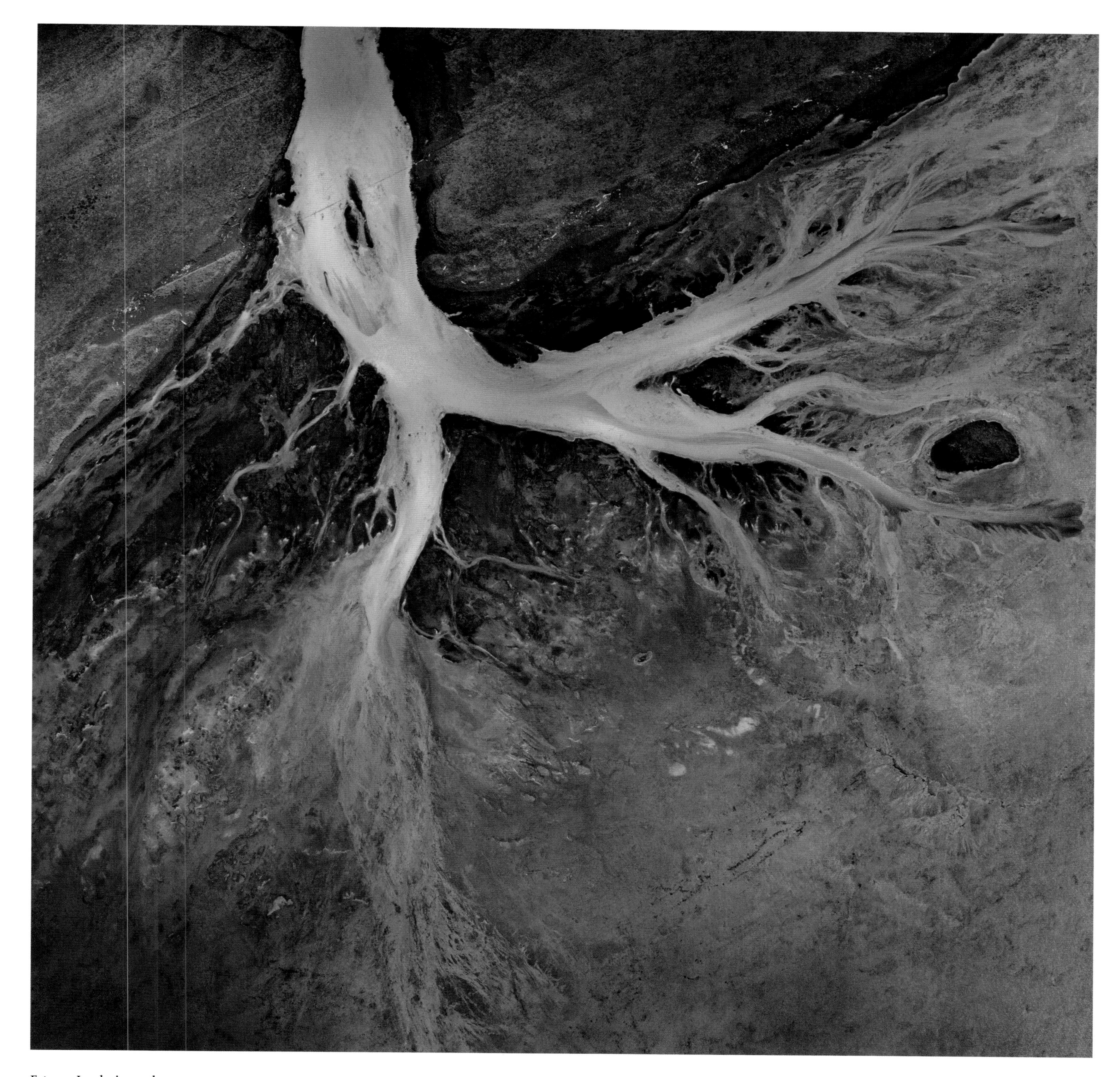

Estuary, Landeyjarsandur
Iceland

Iceland's south coast is bordered by large flat plains of black volcanic sand which extend far into the sea, like Landeyjarsandur. In the tidal zone, rivers carrying vast amounts of meltwater from Iceland's glaciers to the Atlantic ocean branch out into countless arms.

When one of these rivers flows across moorland, its water becomes acidic and can absorb iron minerals out of the black volcanic rock. Wherever these minerals are later deposited, they give the river-bed a reddish yellow colour.

Mývatn
Iceland

The scarred headland in Mývatn, which means 'Lake of Midges', in the highlands of Iceland was created by a stream of lava that flowed over marshland around 2,300 years ago. The glowing molten rock heated water in the ground to the point where it abruptly evaporated. This caused sudden explosions, which in turn created countless small craters. As these small cones have never thrown out lava or ash, they are known as pseudocraters.

The whole of Mývatn is the result of this ancient eruption. The lake was created when one of the many lava flows solidified in front of a depression in the landscape, acting like a dam behind which rain and meltwater collected. It has a surface area of 3.5 square kilometres and a maximum depth of only four metres.

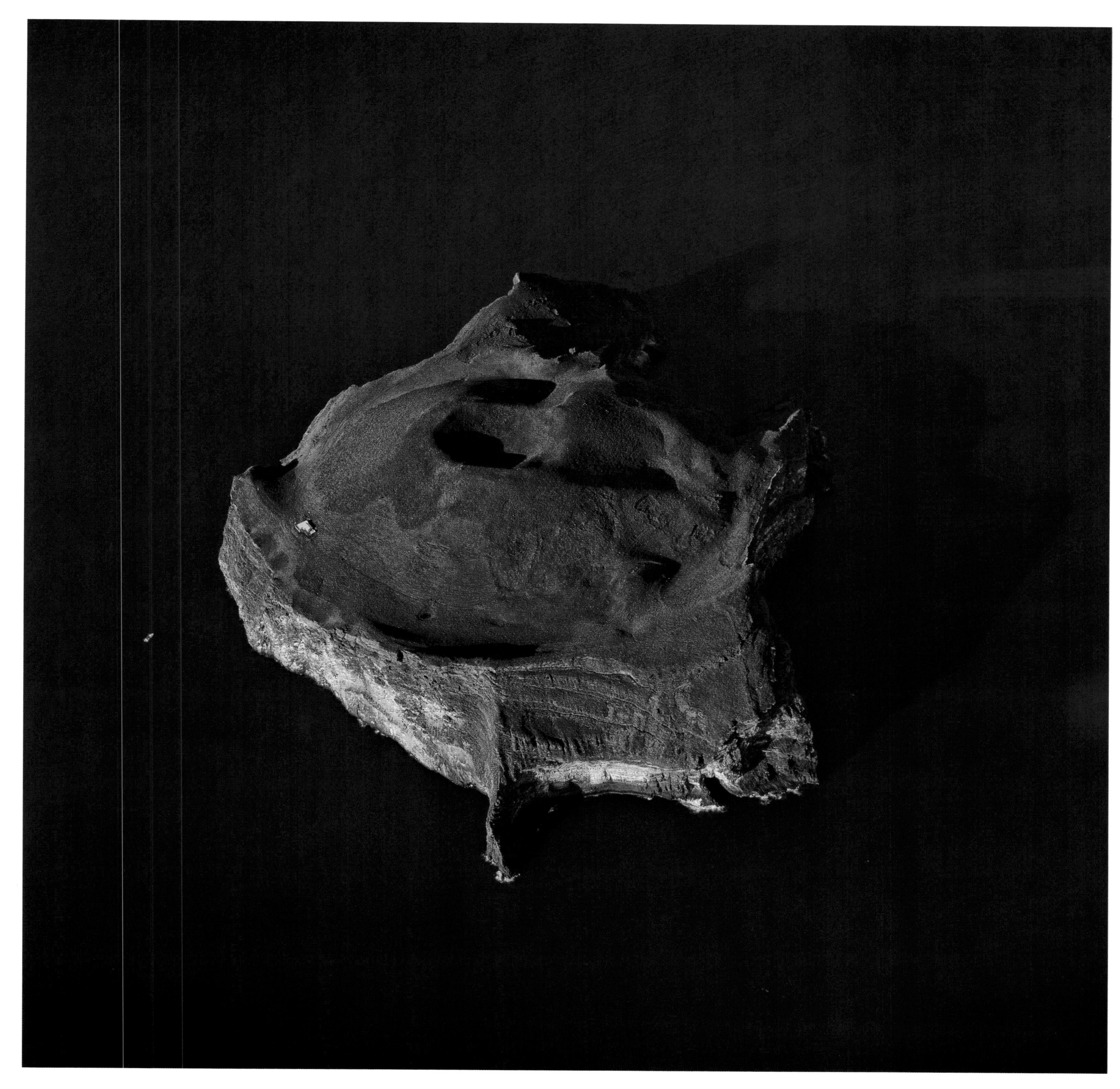

Bjarnarey
Vestmannaeyjar Islands, Iceland

The island of Bjarnarey has an area of about 0.3 square kilometres and rises to 161 metres at its highest point. It is the fourth largest of the 15 Vestmannaeyjar islands off Iceland's south coast, which were created during two or three volcanic periods of the last 8,000 years. They lie along the fracture zone of the Mid-Atlantic Ridge, which runs across Iceland and continues under the sea.

Bjarnarey is only 6,000 years old. Two massive eruptions have occurred in this archipelago in recent years. Between 1963 and 1967, the island of Surtsey rose out of the ocean among spectacular ash explosions. On Heimaey, the largest island, an entire district was buried beneath ejected ash and the port was threatened by a lava flow during a fissure eruption in January 1973.

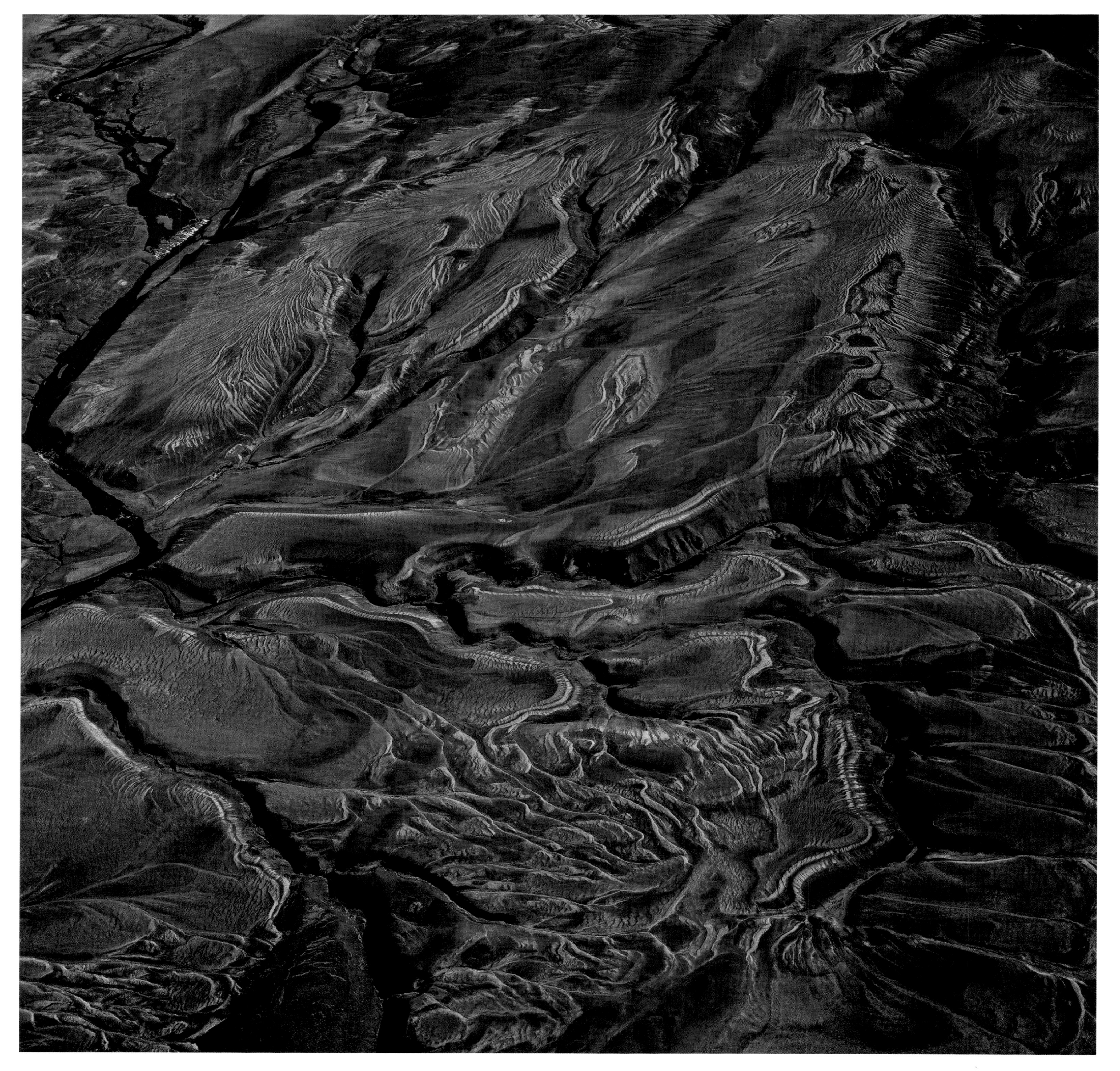

Plain at Hekla volcano
Iceland

Rivers flush out deep channels in the dark desert areas around the base of Hekla, Iceland's most active and most famous volcano, exposing bands of light pumice in the black volcanic deposits – evidence that Hekla is no ordinary fissure volcano, which usually only eject streams of lava during eruptions.

A ridge 27 kilometres long, 2.5 kilometres wide and 1,431 metres high, Hekla is a mixture of fissure volcano and stratovolcano. During its highly explosive eruptions, it spits clouds of ash over 20 kilometres high into the atmosphere, most recently in 1947–8. The entire island was covered in ejected material and ash rained down on Finland. In the Middle Ages, not only Icelanders but people across Europe saw Hekla as the 'Gateway to Hell'.

Reykjanesviti
Iceland

Volcanic soils can resemble chemical factories, particularly in areas where they are very wet and hot. The vapours streaming out of cracks and fissures in the rock often reach temperatures of many hundreds of degrees Celsius, and transport not only sulphur minerals but also metallic compounds from the volcanic bedrock to the surface. As soon as they make contact with the air, they deposit their cargo. Iron oxide, for instance, can colour the ground red, while iron sulphide makes it charcoal grey. Nickeline salts create greenish marks and copper sulphate a bluish colour, while the yellow is either sulphur or iron hydroxide. However, these colours only offer up clues – verifying individual substances requires extensive chemical analysis.

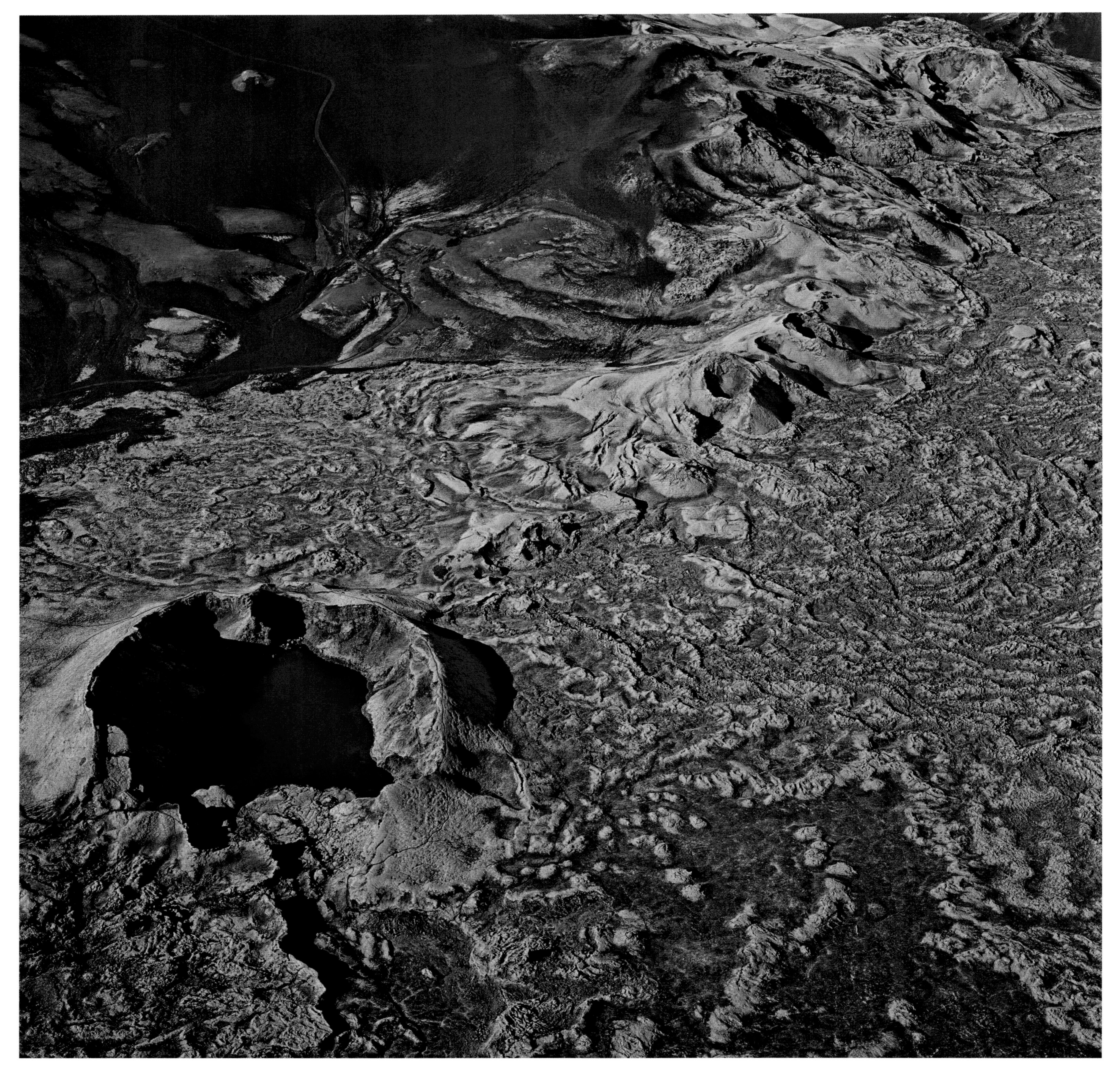

Laki fissure
Iceland

More than 130 craters line up along the 24-kilometre long Laki fissure in the south of Iceland, all of them created during an eruption in 1783. In the worst natural disaster in Iceland's history, between 14 and 15 cubic kilometres of magma came out of the earth here in less than one year. Barely one per cent of it was thrown into the air as ash; the rest flowed out as streams of lava. At times, 8,000 cubic metres of molten rock per second oozed out of the ground, while fountains of lava shot hundreds of metres into the sky.

The lava filled in two river valleys and covered more than 50 square kilometres of land. Poisonous volcanic gases killed livestock, plants died away, and the resulting famine led to 10,000 deaths, a fifth of the population at the time.

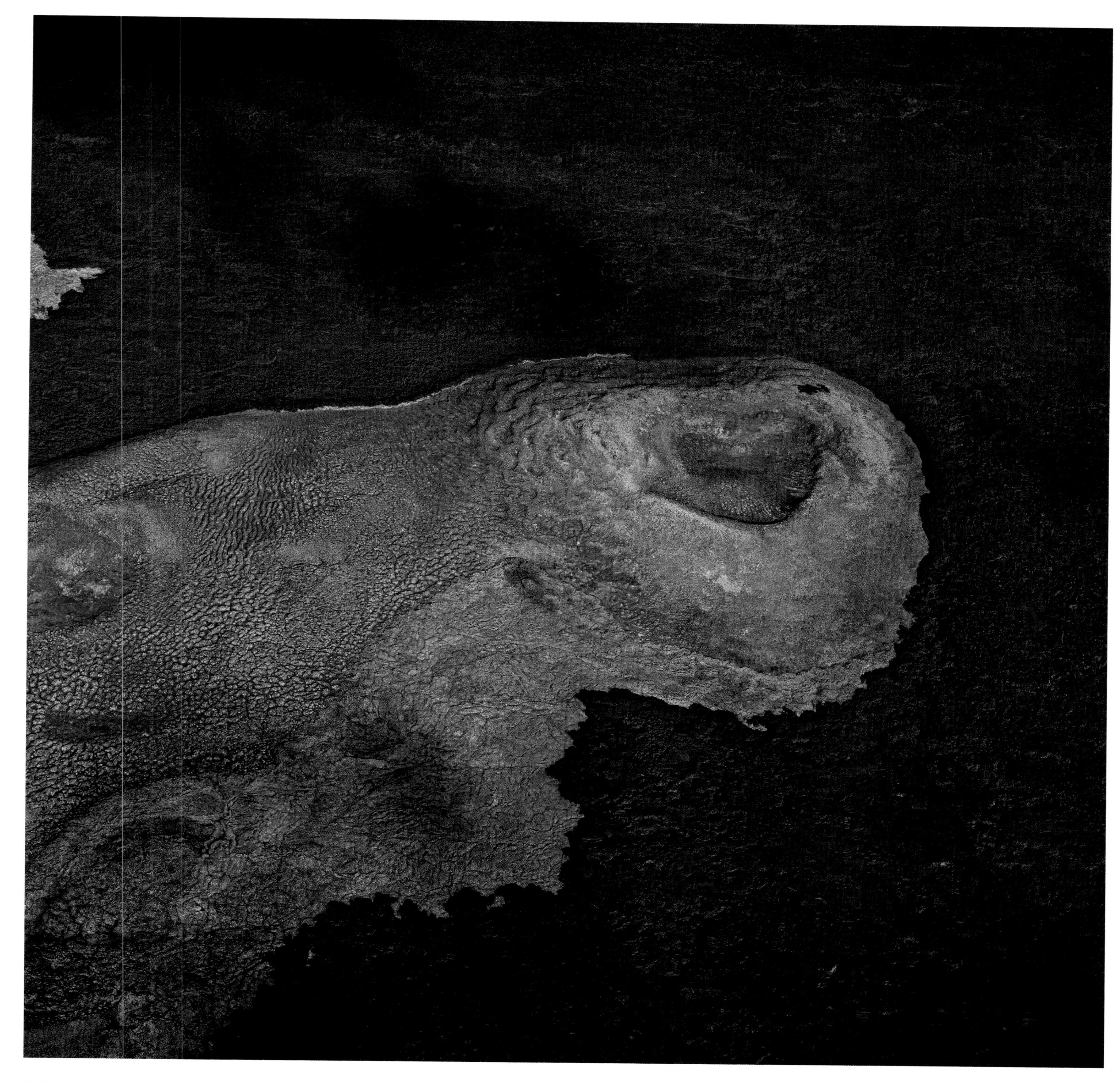

Krafla volcanic region
Iceland

The black lava flows surrounding this old crater cone covered in moss and grass date back to the last great eruption in the Krafla volcanic region in the north of Iceland. This eruption phase lasted nine years, from 1975 to 1984. Flow structures of the fresh lava are still clearly visible, while topsoil has already built up on the surface of the old cone, allowing it to be colonized by plants.

Countless parallel cracks in the earth are distributed over 40 kilometres of the Krafla volcanic area, where the earth's crust expanded by around 10 metres during the last eruption phase.

Bláhver, Hveravellir
Iceland

Bláhver, or 'blue spring', is one of the most beautiful hot springs in Iceland. The water has a temperature of around 90 degrees Celsius, and owes its delicate, light blue colour to tiny drops of silicon dioxide. As the hot, silicon-dioxide-rich water cools, quartz starts to crystallize and build up a hard white crust. This geyserite, as geologists call it, surrounds the entire basin of the spring.

This spring in the Hveravellir geothermal area lies on the Kjölur, one of few gravel roads that crosses the desolate highlands of the island – Hveravellir means 'plain of the hot springs'. There are other spectacular springs, as well as fumaroles and solfataras, in this area, which lies at the foot of Strytur, a flat shield volcano. The ice of the Hofsjökull glacier can be seen on the horizon.

Svartsengi, Reykjanes Peninsula
Iceland

Svartsengi is one of Iceland's high-temperature regions, where sea-water that has seeped into the cracks and pores of the volcanic rock is heated to a temperature of over 100 degrees Celsius. Since 1978 this water has been used to produce electricity and thermal heat by a geothermal power station.

After it has passed through the power plant, the water still has a temperature of 50 degrees Celsius. It flows into a basin in an adjacent lava field, where a luminous blue lake has formed in the strange black volcanic rock. Locals used to bathe in this lake, and the 'Blue Lagoon' has now been turned into a world-famous thermal spa.

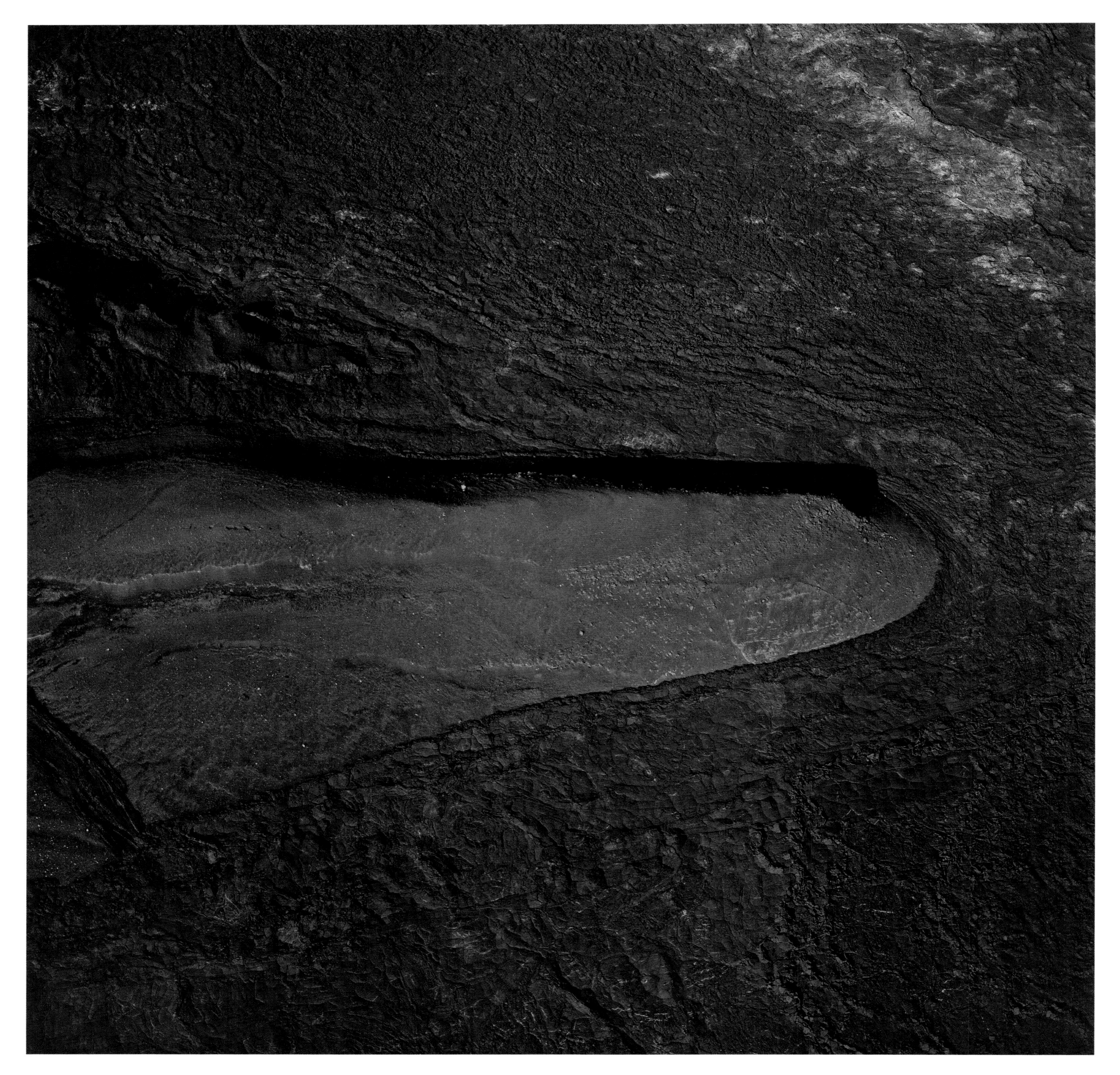

Krafla volcanic region
Iceland

The black lava flows around this elongated red hill date from the large eruption in the Krafla volcanic area. In several phases over a nine-year period ending in 1984, large amounts of lava shot out of fresh fissures, in fountains over 100 metres in height, before turning into gentle streams. The red ridge also consists of volcanic rock, but from an earlier eruption. It has already been smoothed down by weathering and erosion. The intense shade of red indicates that this volcanic rock is rich in iron, as this colour is created when iron in rock oxidizes at high temperatures.

Veidivötn
Iceland

Deserts of ash, rivers, lakes, craters and lava fields shape the young, sparse landscape of Veidivötn in the southern highlands of Iceland, the product of a fissure eruption in 1480. Because magma came into contact with groundwater, the eruptions were accompanied by massive explosions, and a four-kilometre long crater field was created. Today many of the craters are filled with lakes.

Veidivötn is at the end of a fracture zone that runs underneath the ice of nearby Vatnajökull, the largest glacier in Europe, and ends at the subglacial volcano Bárdarbunga. The famous Laki fissure, where Iceland's most devastating volcanic eruption occurred, is just 25 kilometres away.

Brennisteinsalda
Iceland

The 855-metre high dome of Mount Brennisteinsalda and the colourful, barren landscape around it are part of a volcanic massif that was buried beneath a huge glacial sheet during the ice age. Brennisteinsalda means 'sulphur wave', and the rock here has been penetrated by hot, corrosive sulphuric vapours, which have given the stone its yellow colour. After the ice disappeared, rivers carved into the rock.

Around AD 1480, a great, black lava flow, known as Laugahraun, poured out of the mountain's eastern flank. Since then, volcanic activity here has been limited to fumaroles and the Landmannalaugar warm springs at the foot of Brennisteinsalda, a natural spa popular with Icelanders and tourists alike.

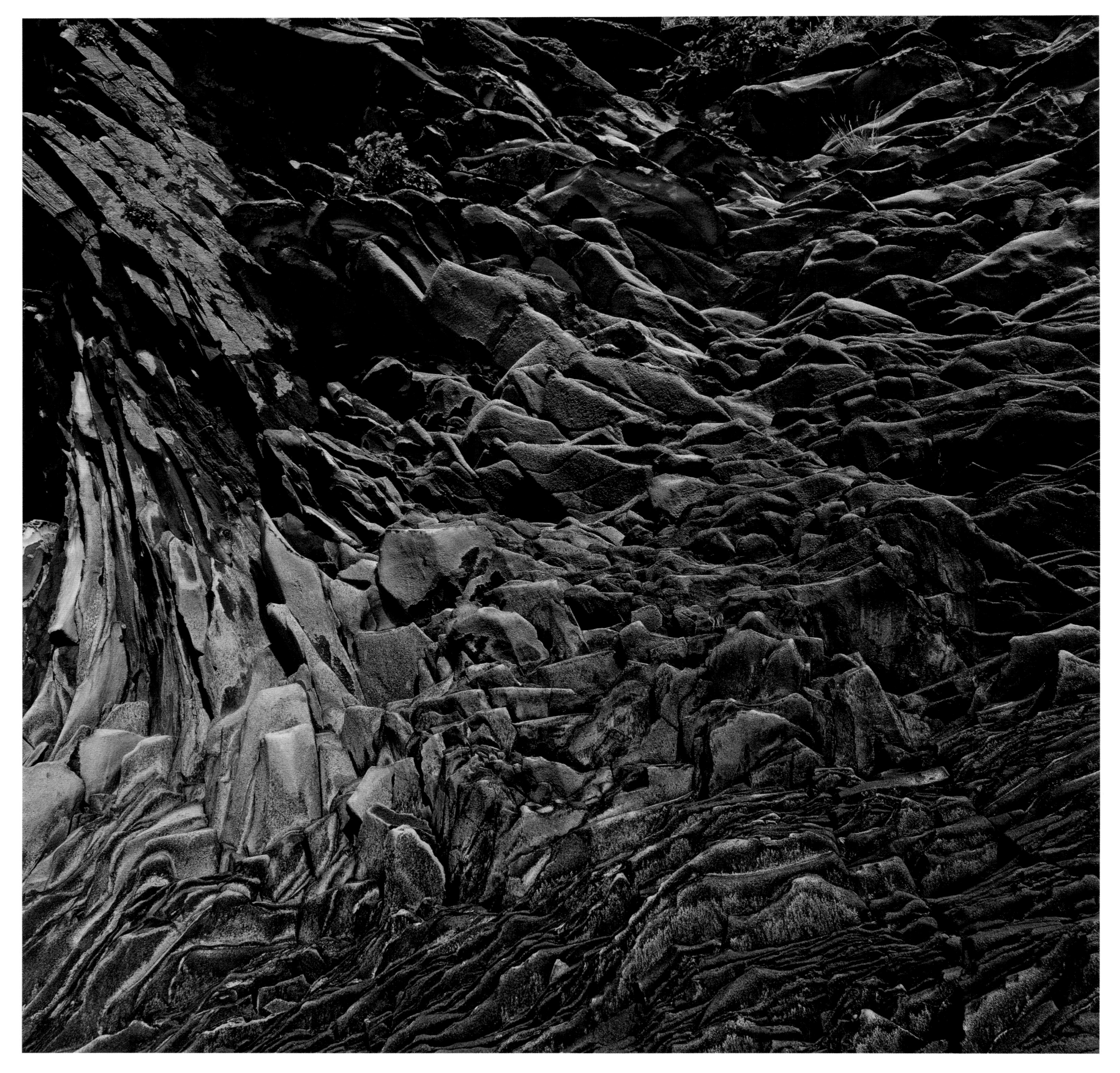

Snæfellsness
Iceland

The constant surging waves of the North Atlantic reveal the layered plate-like basalt structures created by an ancient lava flow from the legendary Snæfellsness volcano. At 1,448 metres, it is the highest peak in the chain of volcanoes on the Snæfellsness peninsula in western Iceland, which extends 80 kilometres into the North Atlantic. It last erupted around AD 900, an event documented in Iceland's *Landnámabók*, or book of settlement.

In his novel *Journey to the Centre of the Earth*, Jules Verne (1828–1905) had Professor Lidenbrock climb through Snæfellsjökull's vent to reach the interior of the planet. The naturalist and his companions were later spat out of the crater of Stromboli in southern Italy.

Mælifellsandur
Iceland

Bright green moss (*mniobryum albicans glacial*) has colonized a hill in the middle of Mælifellsandur, a black desert of lava and volcanic ash in the south of Iceland. The hill is all that remains of a once active cinder cone, ground down by ice from the nearby Mydralsjökull glacier. The red patches are rich in iron oxide. Mælifellsandur is a dark volcanic plain in the north of the Mydralsjökull glacier, the fourth largest of Iceland's 13 major glaciers.

Svartifoss
Iceland

When large streams of lava gradually cool down, they often solidify in the form of columns. The molten rock contracts as it cools and cracks into polygonal sections, creating countless, tightly packed, three- to eight-sided stone columns. In Iceland, these natural columns are primarily found along rivers and around waterfalls and cliffs. Made up of basalt lava, the columns at the Svartifoss waterfall in the Skaftafell National Park in the south of the island are among the most beautiful in Iceland.

Skeidararjökull
Iceland

The characteristic patterns made by volcanic ash on Iceland's glaciers, like these on Skeidararjökull, a branch of the great Vatnajökull glacier in the south of the island, are the result of a complex process in which sun, wind and snow play important roles. Iceland's many volcanoes, which erupt every few years, supply the black ash, which rains down from eruption clouds, while the wind blows dark sand and dust from nearby volcanic deserts onto the ice. Snow falls onto this dark carpet of ash, and white covers black, layer upon layer. Because ice flows, this black and white layer cake slowly moves forwards, causing the layers to distort, compress and spread out, while wind and sun erode and cut up the surface.

Námafjall geothermal area
Iceland

The ground at the foot of the Námafjall ridge is hot, as are the volcanic gases and vapours that slowly hiss out of its holes and fissures – among them hydrogen sulphide gas, spreading its unpleasant smell of rotten eggs. As usual, the hydrogen sulphide turns into sulphuric acid the moment it reaches the surface and reacts with the oxygen in the air. Over time, this corrosive substance erodes the hard volcanic rock, turning it into soft yellow clay. Pure sulphur is also deposited in large quantities around solfataras and fumaroles. In previous centuries it was mined and exported for use in the production of gunpowder.

Assal Rift, Djibouti
Lake Assal, a salt lake, lies on an active spreading zone – a rift. The earth's crust regularly tears open here, great slabs of rock are staggered like steps and lava pushes up out of cracks that are often hundreds of metres long. This black lava flow streamed out during a fissure eruption in 1978.

One of the most interesting volcanic regions on earth is found on the African continent: the East African Rift, part of the Great Rift Valley. It begins in the Afar Depression, on the southern edge of the Red Sea, then runs south for 4,000 kilometres through the deserts, savannahs and forests of East Africa before disappearing between Lake Malawi and the coast of Mozambique.

The earth is expanding here. Magma bubbles up through fissures and cracks in the rock. There are more than one hundred active and countless extinct volcanoes along the fracture zone, including Mount Kilimanjaro, Africa's highest mountain, and Ol Doinyo Lengai, the only volcano on earth to produce 'cold' lava.

According to scientists, Africa could split in two along the East African Rift at some point in the next few million years, with a new ocean being created between the two sections as they drift apart.

Africa

Crater fields
Marsabit volcano, Kenya

There are more than 200 craters on the flat flanks of Marsabit, a volcano that rises about 1,000 metres above the Chalbi Desert in northern Kenya like a green island. This mighty shield volcano lies on the furthest edge of the East African Rift, and is around five million years old. The small cones on its ridge emerged about 500,000 years ago, when Marsabit began spitting lava and ash again after a long dormant period, allowing molten material to break through the old volcano's slopes in many different places.

The volcano's rock acts like a sponge. It stores the water that falls from the sky during the short rainy season in dry northern Kenya, making agriculture possible on the land.

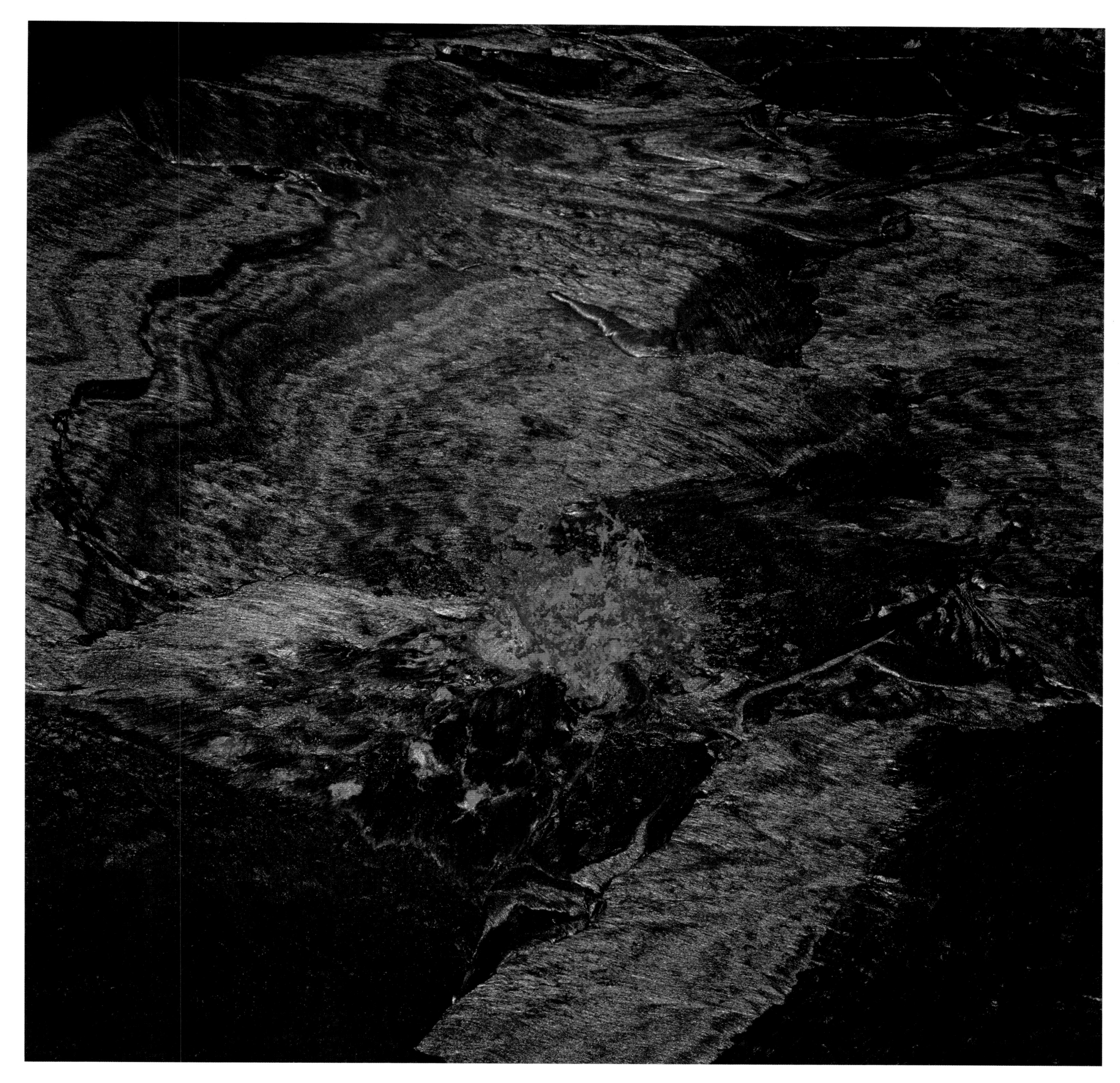

Erta Ale
Danakil Desert, Ethiopia

The Erta Ale volcano harbours a very rare volcanic phenomenon – a lava lake that has been bubbling in one of its two craters for over 90 years. It is covered by a thin, black solidified crust. The molten material beneath it is in constant motion, like hot porridge in a saucepan. The black skin rises and falls, contracting in one place and tearing apart in another. From time to time lava gushes out of these openings, creating fountains that can reach heights of 15 to 20 metres.

Erta Ale is the most active volcano in the Afar Depression on the northern edge of the East African Rift. The name Erta Ale comes from the language of the Afar, the nomadic people who live here, and means 'smoking mountain'.

Nyamulagira
Virunga Mountains, Democratic Republic of Congo

Nyamulagira is currently Africa's most active volcano, but its spectacular eruptions receive very little attention due to its remote location in the dense forests in the east of the Democratic Republic of Congo, near rebel-controlled areas on the Rwandan border.

The volcano erupts around every two years. In May 2004, fountains of lava shot out of a fresh, several-hundred-metre long crack in the the flank of the 3,085-metre high shield volcano, finally concentrating at two vents where glowing molten rock was continuously thrown up to 100 metres in the air. Since 1882, Nyamulagira has erupted more than 40 times.

Nyamulagira
Virunga Mountains, Democratic Republic of Congo

Over the course of an intense eruption that began in May 2004 and lasted for about 30 days, lava was mainly expelled from just two openings, where two 50-metre high craters grew. Towards the end of the eruption, the pulsating molten rock in the north crater (left in the photograph) regularly grew into huge, 20-metre wide bubbles that lasted for fractions of a second before bursting with a deafening bang. According to official reports, no one was injured in the surrounding area, but large expanses of tropical forest were destroyed.

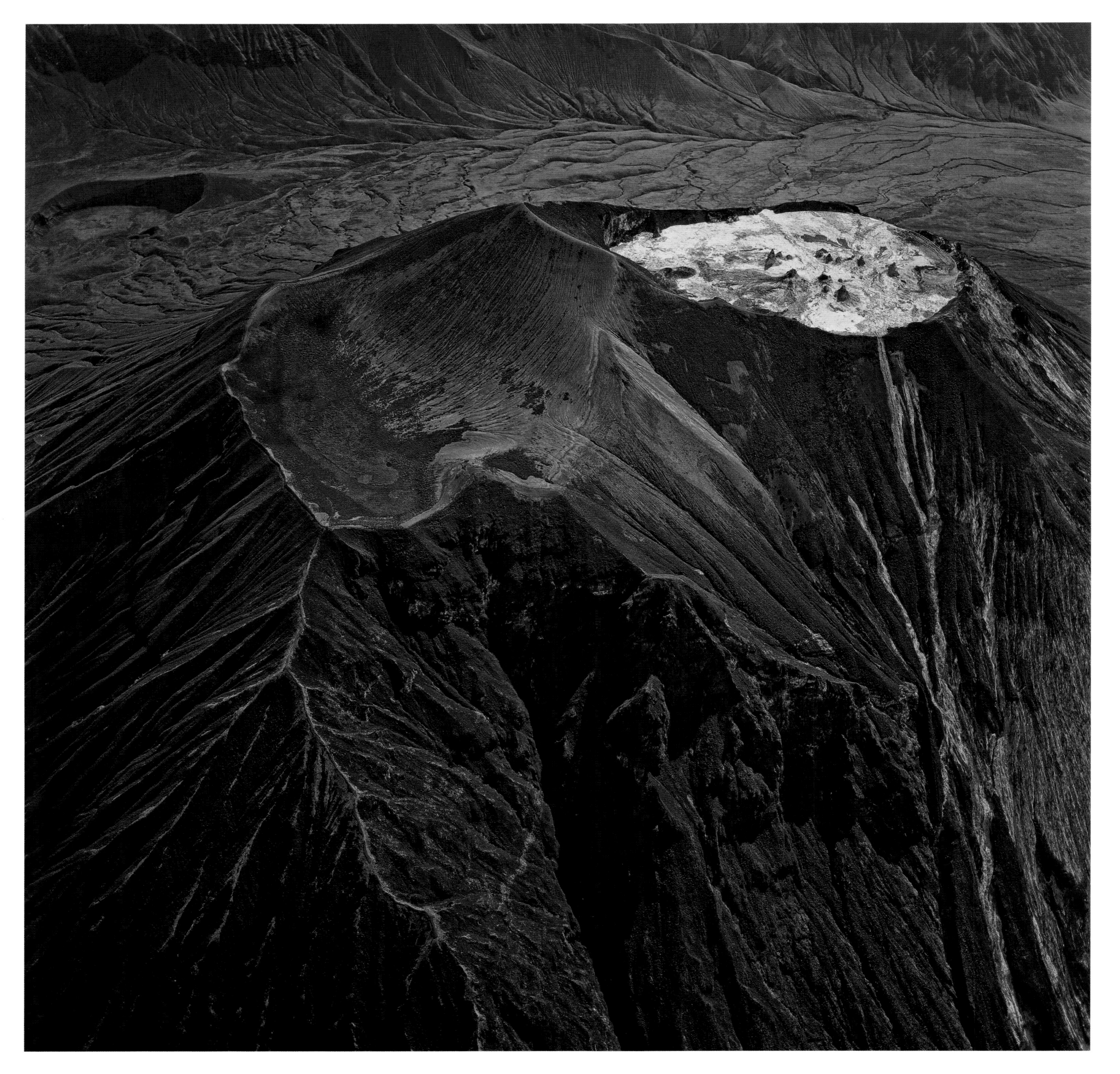

Ol Doinyo Lengai
Tanzania

The lava of Ol Doinyo Lengai is unique. It contains sodium carbonates as well as potassium compounds, and can therefore still be fluid at temperatures of around 550 degrees Celsius – the normal temperature of flowing volcanic lava is 1,000 degrees Celsius. It bubbles up like viscous, black oil from chimney-like eruption cones known as hornitos that rise up from the volcano's flat crater floor. As the lava solidifies, it reacts with moisture in the air, its surface turning first light brown and finally white.

Ol Doinyo Lengai is an almost symmetrical stratovolcano, towering 2,000 metres above the savannah in the East African Rift. The Masai people who live at its foot gave the volcano its name – Ol Doinyo Lengai means 'God's mountain'.

Guyot
Awash River flood-plain, Ethiopia

There are a number of unusually flattened volcanic cones in the large Afar Depression at the northern end of the East African Rift, such as this one rising up over the flood-plain of the Awash River. This form is typical of guyots, which are underwater volcanoes. They serve as proof that the Afar Depression was once a branch of the Red Sea. The French geologist Haroun Tazieff (1914–98), the first scientist to explore the remote deserts of the Afar Depression, discovered the remains of coral reefs on the flanks of these volcanic stumps – evidence that they had been created under water.

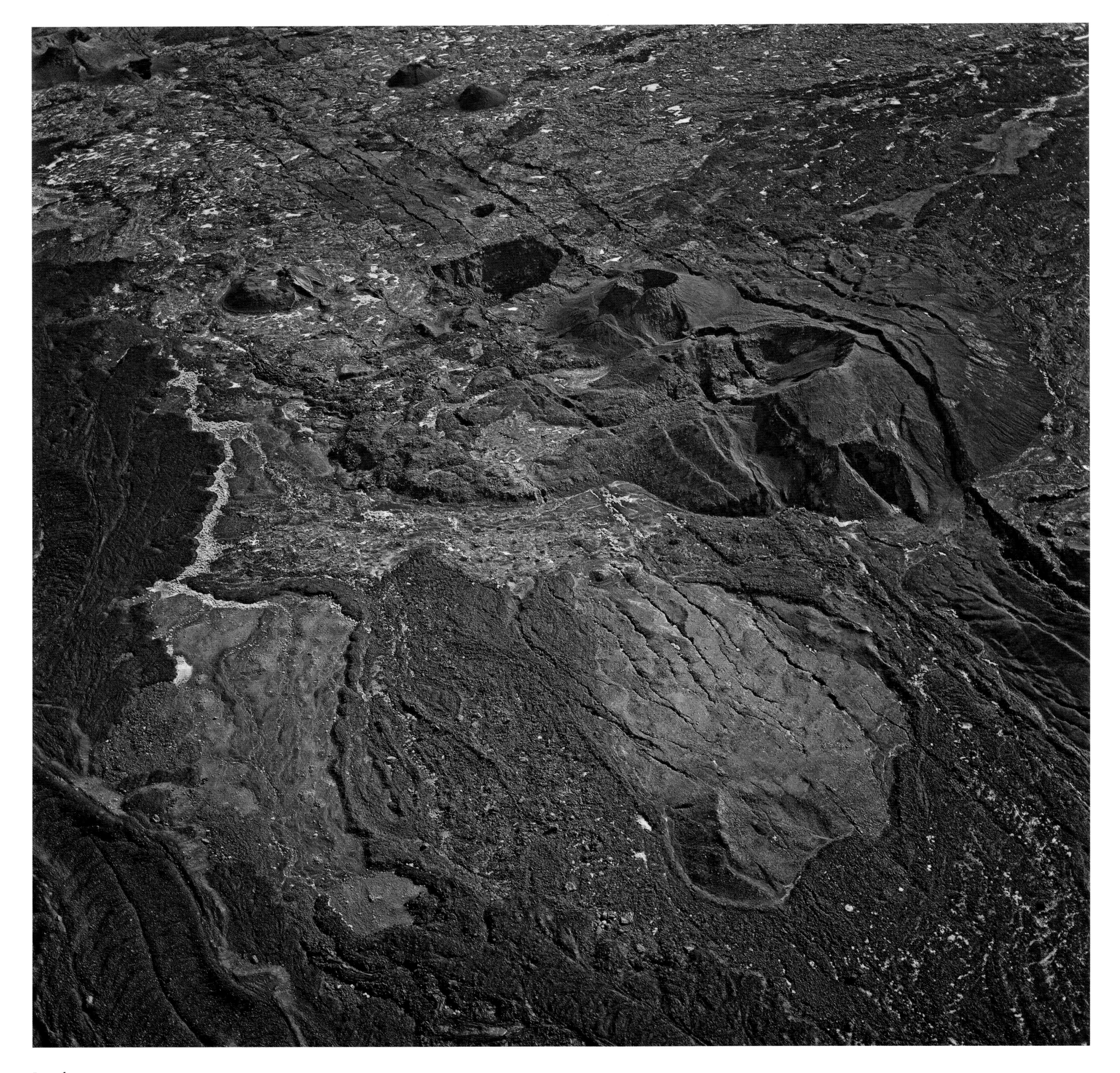

Lava desert
Danakil Depression, Ethiopia

Clusters of parallel fissures, hundreds of metres long, run across lava fields and craters in the Danakil Desert in the northern Afar Triangle. They are evidence that the earth is spreading here, at an average rate of one to two centimetres per year according to satellite measurements. This process is not continual, but takes place in surges, as scientists were able to witness in autumn 2005. Within a fortnight, the ground ripped open over a length of about 60 kilometres during around 150 perceptible earthquakes. Some fissures widened by eight metres.

These barren, remote deserts clearly reveal the events taking place in the world's rift zones – magma pushes up from the depths, then stretches the earth's crust until it rips and finally breaks into large plates, which drift apart.

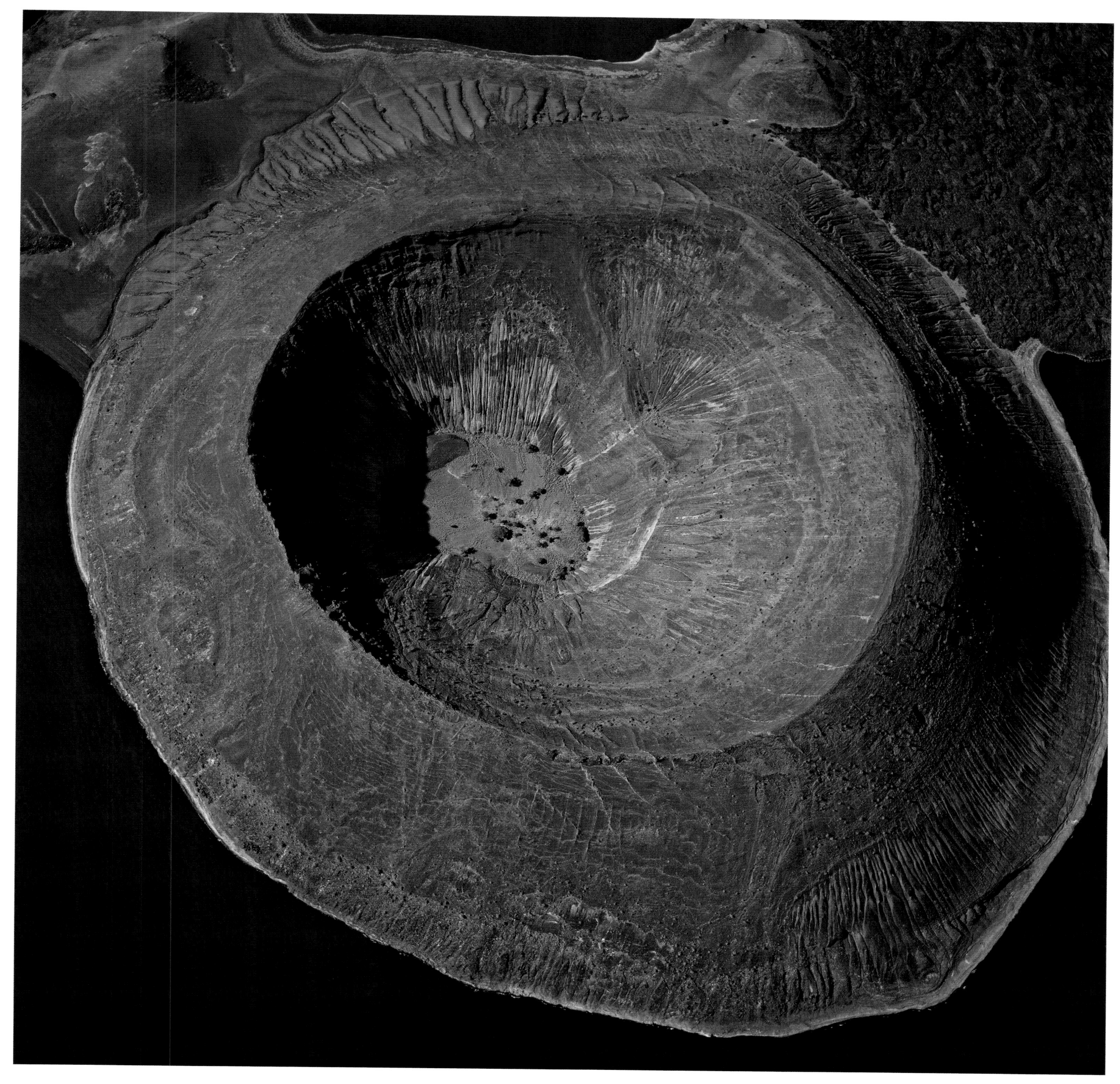

Nabuyaton crater
Lake Turkana, Kenya

Nabuyaton is a huge crater on the southern shore of Lake Turkana, one of the long, narrow lakes in the East African Rift. It belongs to a volcanic complex known as the Barrier, which separates Lake Turkana from the basin of a dried-up lake to the south.

The first European to reach this remote African desert region was the Hungarian count Sámuel Teleki (1845–1916), whose expedition reached the southern end of Lake Turkana in March 1888. He discovered the crater cones of the great Barrier shield volcano, and a cone near Nabuyaton is named after him.

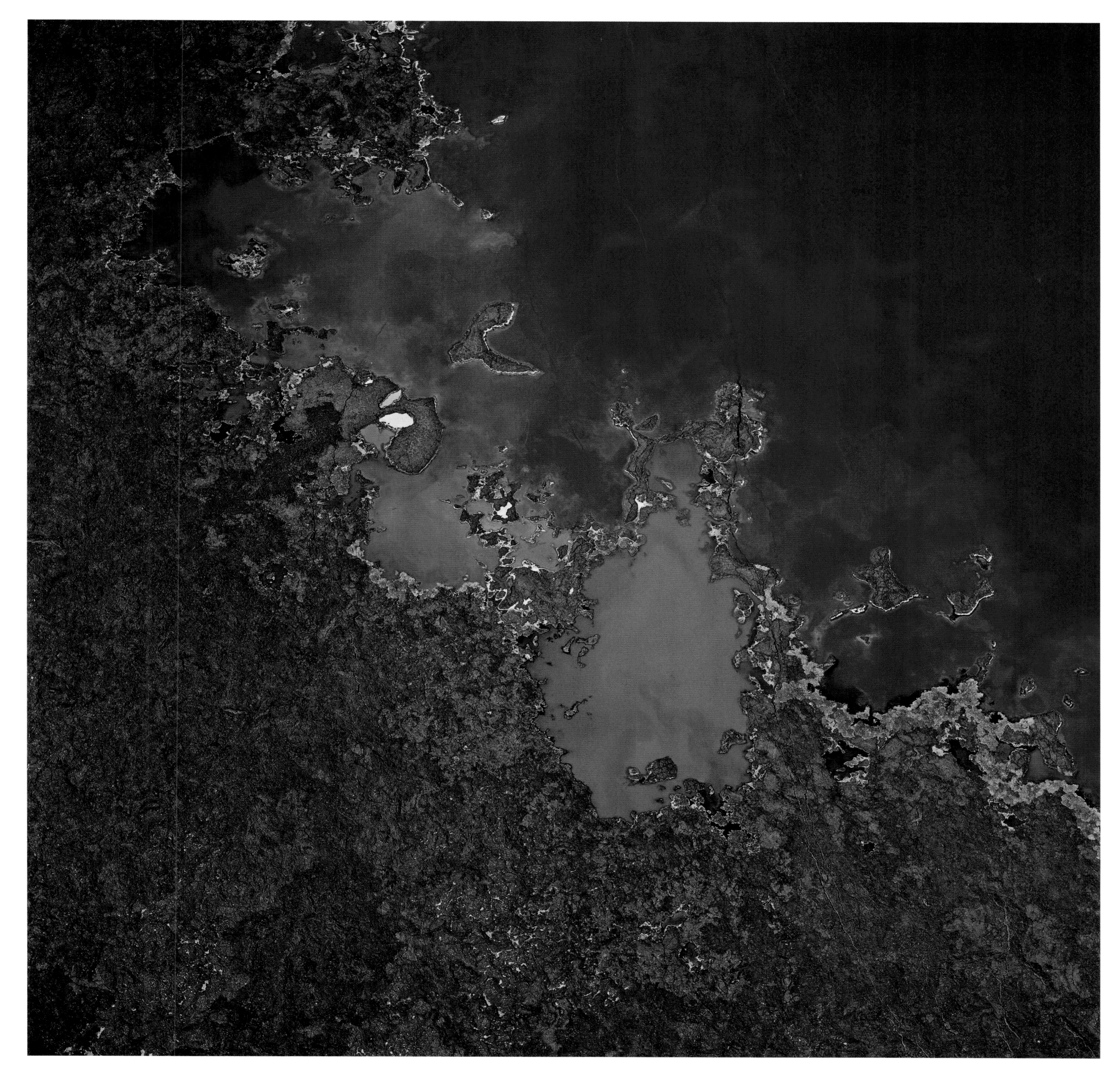

Geothermal area on the shores of Lac Assal
Djibouti

The geothermal springs that bubble up out of the dark volcanic rock on the southern shores of Lac Assal reach temperatures of up to 80 degrees Celsius. The areas where hot water flows out are coloured green and orange by algae and bacteria that can not only withstand the high temperatures, but also the water's high salinity. There are up to 348 grams of salt in a litre of lake water, which is a salt content of more than 30 per cent. In comparison, sea-water has an average salinity of only 3.5 per cent, a tenth of that of the lake.

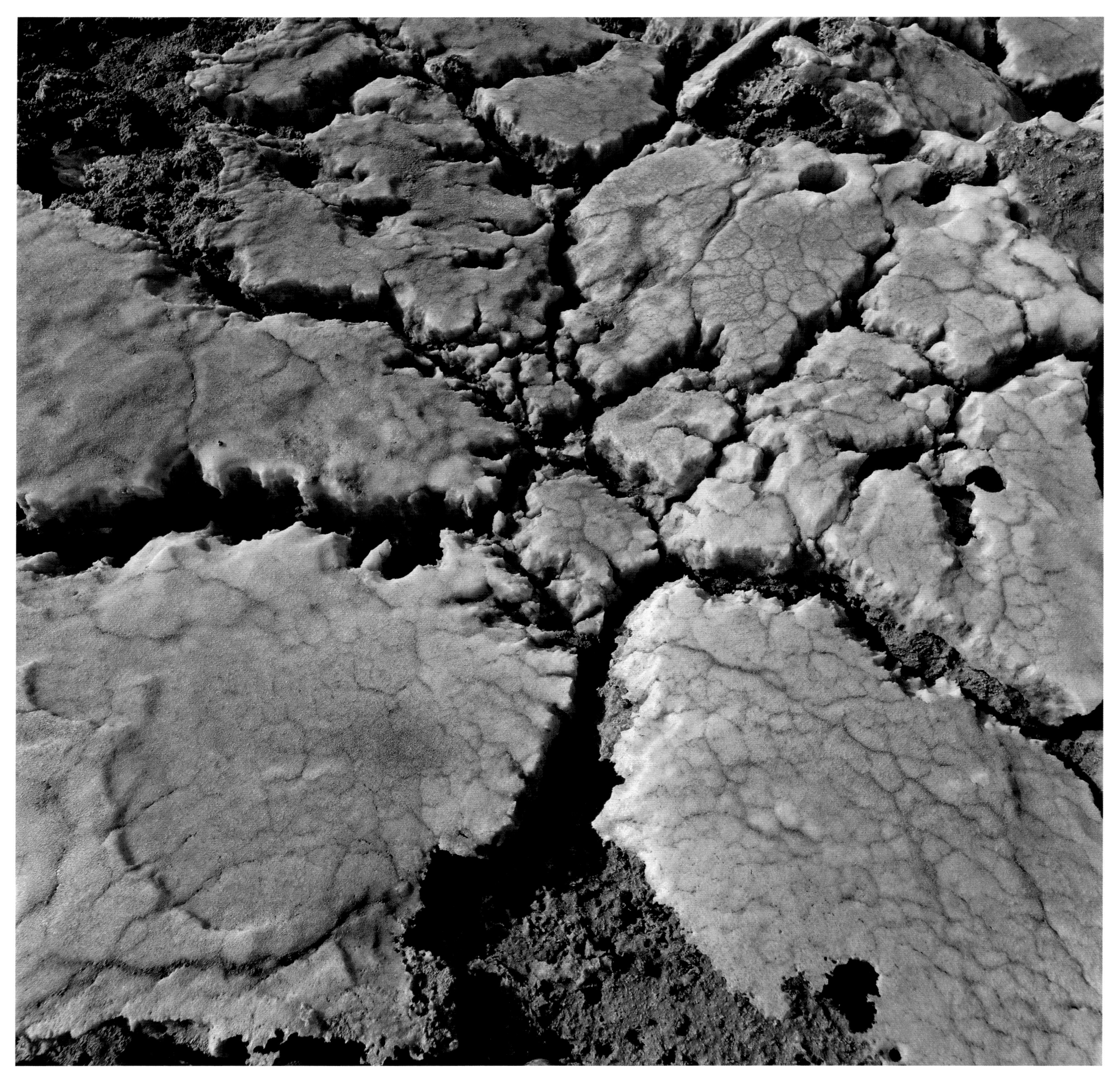

Salt crust
Dallol, Ethiopia

A thick salt crust covers the desert floor in Dallol, a geothermal region in the north of the Afar Depression, near the border between Ethiopia and Eritrea. Like guyots, those flattened volcanic cones, the salt in the ground indicates that the Afar Depression was once a shallow branch of the Red Sea. During a period between 65,000 and 125,000 years ago, it repeatedly dried out only to be flooded again, and salt was deposited here as result. This salty crust is more than 1,000 metres thick. Hot springs wash the salt out of the ground, and it is deposited around the spring pools, cracking into a web-like pattern as it dries. The salt crust is stained yellow by sulphur and iron compounds.

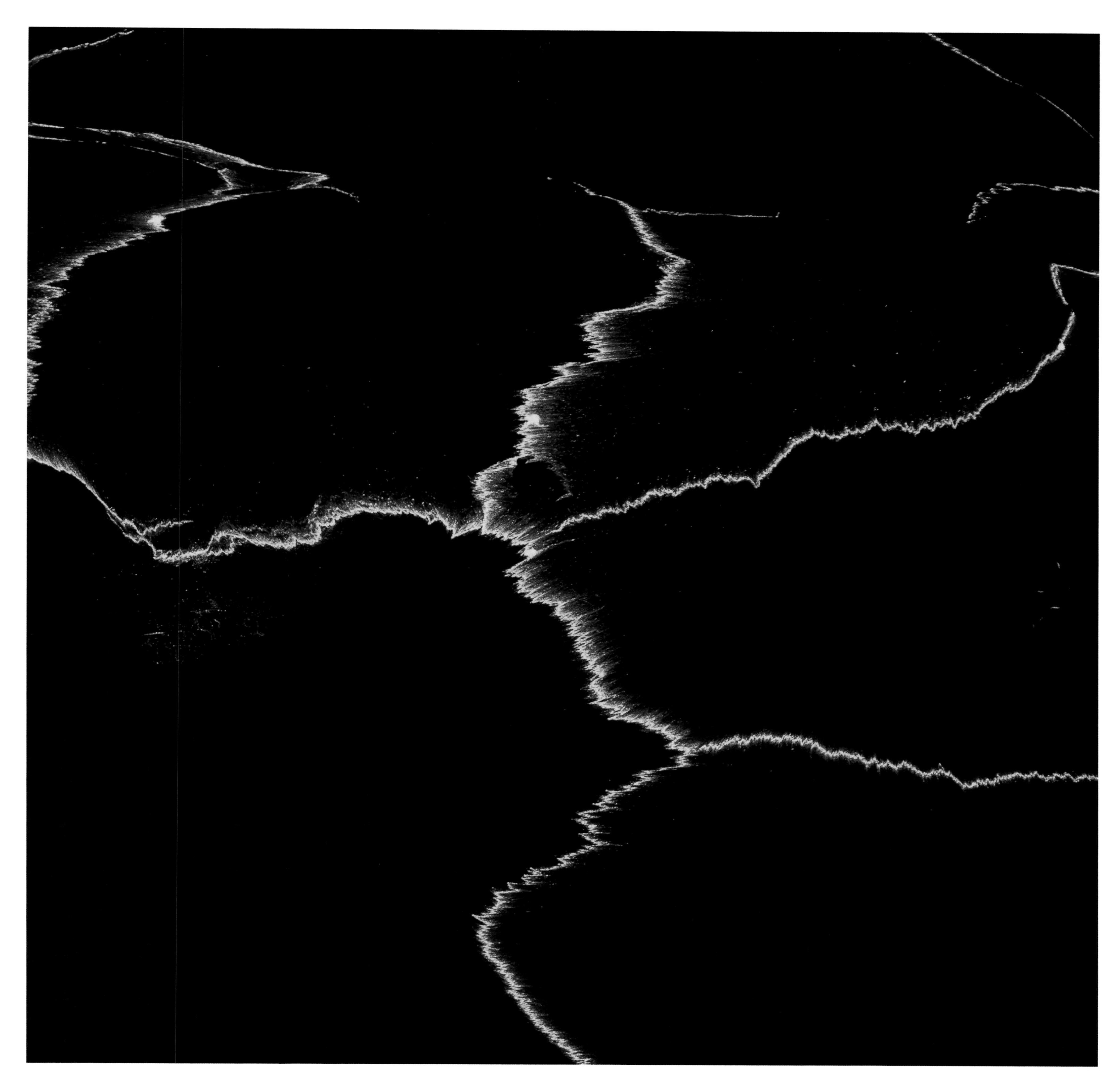

Erta Ale
Danakil Desert, Ethiopia

A network of glowing fissures cuts through the dark skin covering the lava lake in the crater of Erta Ale, a shield volcano. The glowing molten rock has a temperature of around 1,200 degrees Celsius. On the surface it cools to around 500 degrees Celsius and forms a thin, flexible crust.

Rips tear the delicate crust into segments that pull apart or dive below one another. Scientists regard the movements in Erta Ale's sea of lava as a model for the earth's plate tectonics: the process shows in miniature what takes place on the earth's surface, where the hard plates of the lithosphere float on the soft asthenosphere, drift apart and collide.

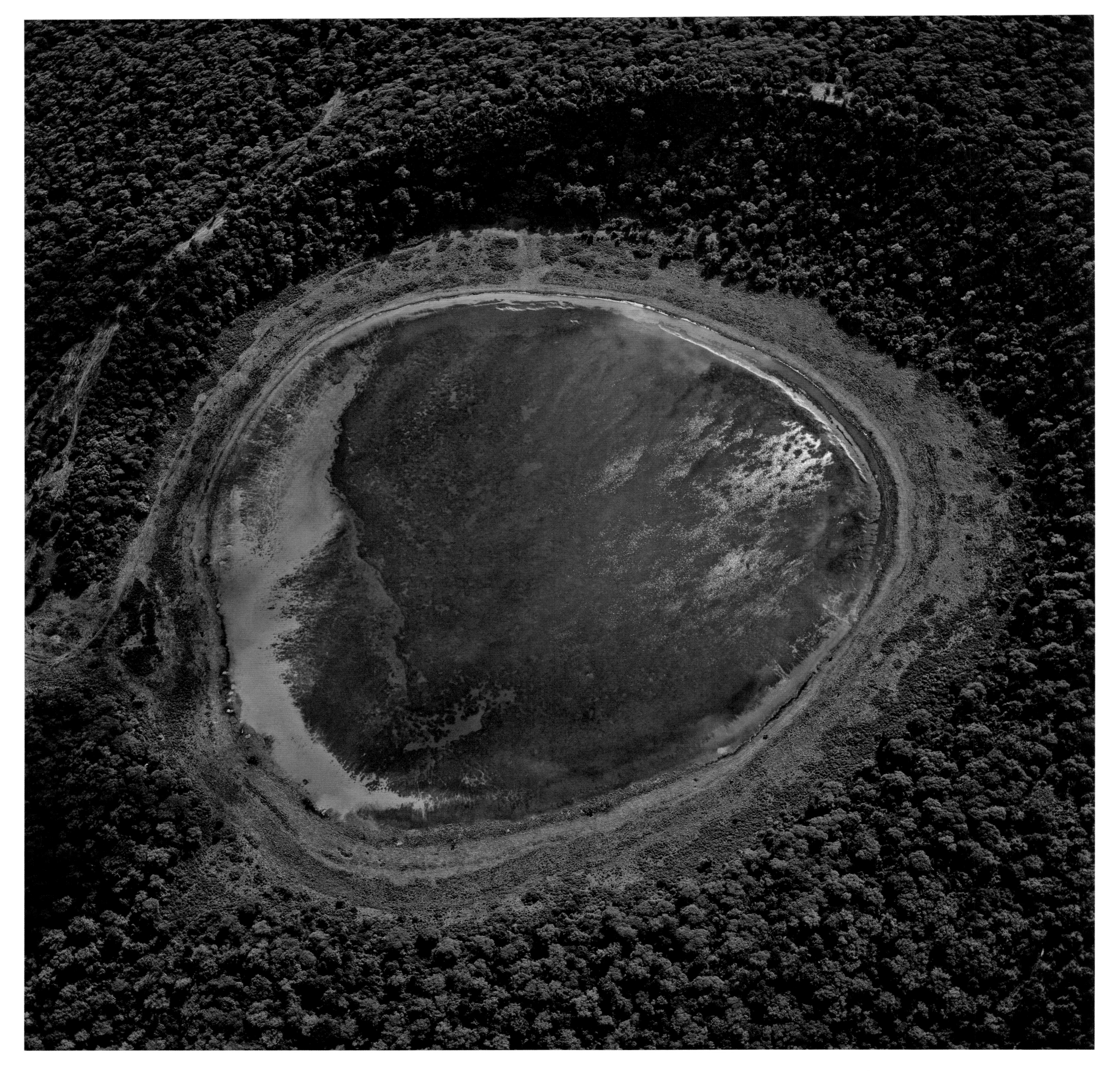

Lake Paradise
Marsabit volcano, Kenya

The almost circular Lake Paradise is a crater lake. Its waters fill the eruption crater of one of the approximately 180 cinder cones created during the flank eruptions of Marsabit, a huge shield volcano in northern Kenya. It is so densely wooded that the shape of the cone is almost impossible to make out.

Lake Paradise lies in the Marsabit National Park and Reserve. Thanks to the herds of elephants that sometimes use it as a watering hole, it is a popular tourist destination.

Maar
Marsabit volcano, Kenya

Like Lake Paradise (opposite), this crater is an eruption opening on the shallow flanks of the large Marsabit shield volcano. One of 22 maars among the Marsabit volcano's 200 parasitic cones, this large hole has a diameter of around 500 metres. It barely rises up over the surrounding area, however, appearing more like the impact crater of a meteorite.

Maars are usually the result of a single, very large explosion. They are created when magma comes into contact with groundwater in the lower layers of a volcano, which then abruptly evaporates and ruptures the earth's surface. Grasses and bushes grow on the moist floor of this maar, although its walls are dry. The largest maars in Marsabit have a diameter of 2.5 kilometres.

Dala Fila
Danakil Desert, Ethiopia

Dala Fila's steep, pastel-coloured cone rises 300 metres over the lava fields of the Danakil Desert in the Afar Depression. The molten rock that wells up from its central vent during eruptions is not as fluid as that in the lava lake of its neighbour, Erta Ale. The magma of Dala Fila is rich in silica and therefore very viscous, so this volcano's eruptions are generally highly explosive. Alternating layers of ash and lava flow build up its steep flanks, and it is therefore classified as a stratovolcano or composite volcano. It last erupted in November 2008 and the gas cloud over its crater is a sign that it is only sleeping.

Ghoubbet al Kharab
Djibouti

The long, narrow bay in Djibouti's Gulf of Tadjoura is called Ghoubbet al Kharab, or 'the devil's gorge'. Black lava fields, crisscrossed with fissures and strewn with cinder cones, meet crystal-clear sea-water in which corals thrive.

The bay lies directly above the Assal Rift, where the earth's crust spreads apart by one or two centimetres a year, so that new fissures, out of which lava flows, are constantly ripping open. The ash thrown out during eruptions creates crater cones up to 100 metres high. The most recent eruption occurred in 1978. During an earthquake, a 1.8-metre wide, 500-metre long rip opened up, out of which shot ash clouds and lava fountains. The eruption ended only 10 days later, after 25 more fissures had opened up parallel to the first.

Crater wall, Erta Ale
Danakil Desert, Ethiopia

The height of Erta Ale volcano's brittle, vertical crater wall changes depending on the level of the bubbling lava lake in its crater. At the time this photograph was taken, it towered about 80 metres above the surface of the lake. The lava fountains that shot out of the glowing cracks in the black skin of cooling lava gushed to heights of up to 20 metres for a few minutes before they died away. Sometimes the lava level is so high that the molten matter spills over the crater wall and flows into Erta Ale's caldera in wide, glowing streams.

A shield volcano, Erta Ale is 613 metres high, 70 kilometres long and 50 kilometres wide. Its shape and the chemical composition of its thin, runny lava resemble those of the underwater volcanoes along mid-ocean ridges.

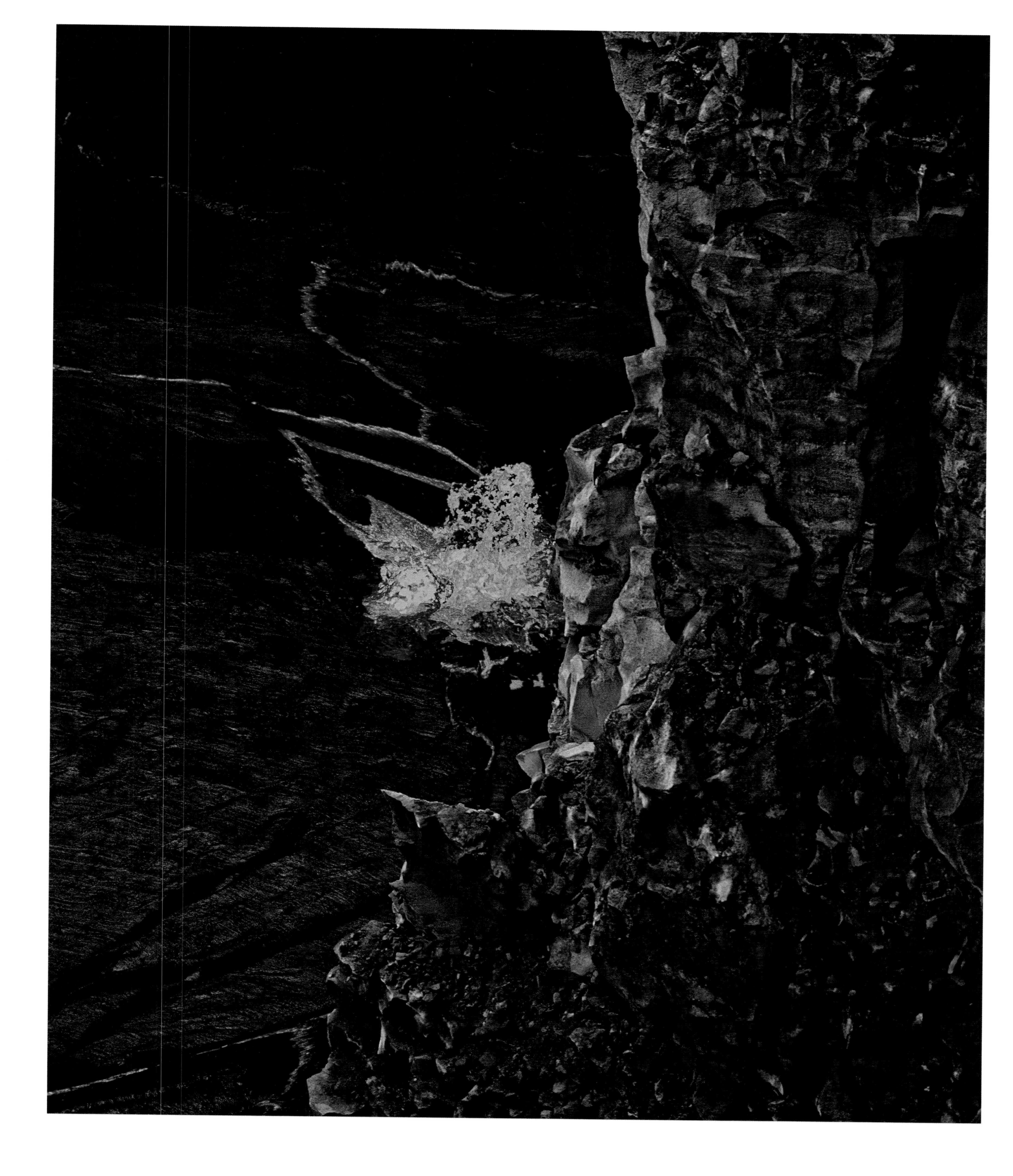

Dallol
Ethiopia

The geothermal area of Dallol in the Danakil Desert is one of the strangest landscapes in the East African Rift. Its lowest point lies 126 metres below sea level, making it the second lowest-lying place in the Afar Depression after the surface of Lake Assal in Djibouti, which lies around 150 metres below sea level.

The scattered hot springs that bubble up from the ground in Dallol are very clear and an intense shade of green. The water is enriched with sulphur and iron compounds and is extremely salty, because it flows through the salt rock that was deposited here when the area was still flooded by the Red Sea. The salt crust that covers large areas of Dallol is furrowed with erosion trenches.

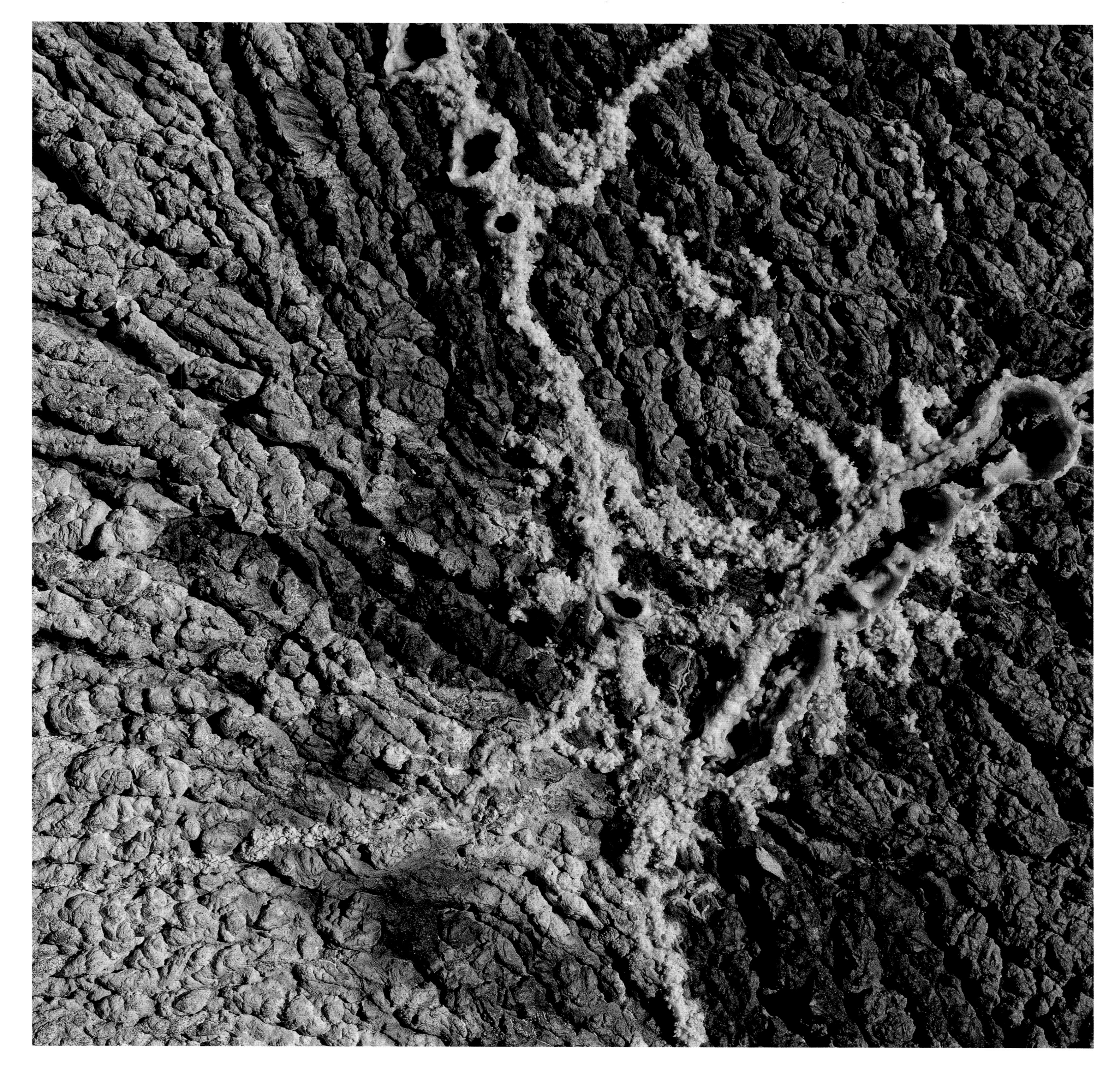

Lava crust, Ol Doinyo Lengai
Tanzania

The lava that flows out of vents in the crater of Ol Doinyo Lengai, the holy mountain of the Masai, solidifies on the surface in twisted, rope-like strands just a few centimetres thick. In some places, the still-hot crust tears open. Gas streams out of these fissures, depositing mineral salts.

Lava usually flows out of a volcano's crater glowing red, turning black when it solidifies, but the lava of Ol Doinyo Lengai emerges at a temperature of only 500 degrees Celsius, already black but still very fluid. Once cooled, it turns brown and brittle after a few days, before the surface finally turns powdery and white. This change is due to the high levels of sodium carbonate, trona and potassium compounds in this volcano's lava, which react with moisture in the air.

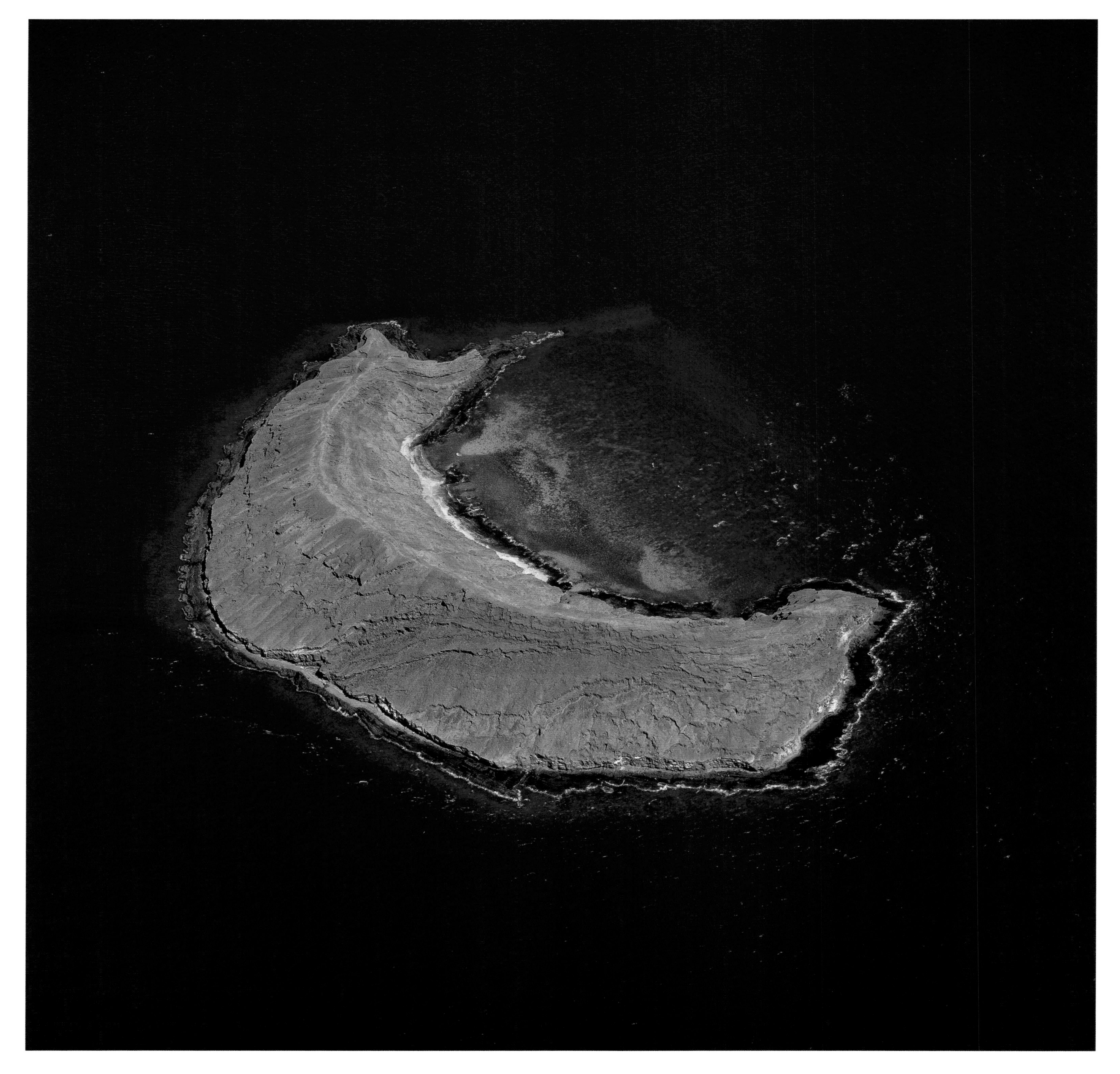

Sawabi Islands
Djibouti

A crescent-shaped island is all that remains of an extinct volcano that is gradually being eroded by the waves, and whose crater has been colonized by corals. This curved mountain ridge is one of the six Sawabi Islands, which lie in a row off the coast of Djibouti in the Bab al Mandab straits. This waterway between Djibouti in Africa and Yemen on the Arabian Peninsula connects the Red Sea to the Gulf of Aden. All six Sawabi Islands are the relics of extinct volcanoes. Together with the Ras Siyan peninsula on the coast of Djibouti, on which the dark red remains of a crater rises up over white sand, they are known as the Sept Frères, or Seven Brothers.

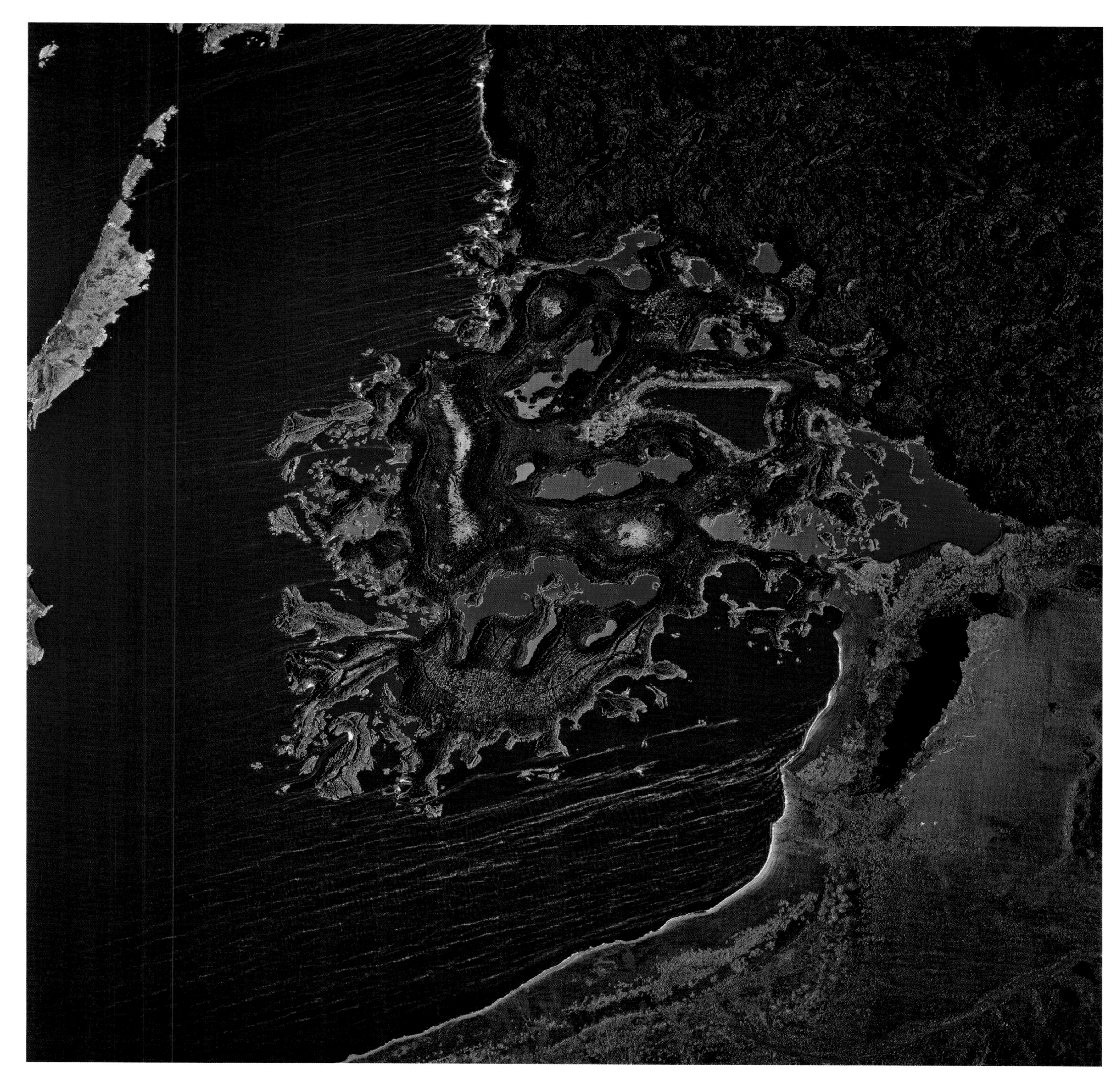

Lake Turkana
Kenya

Thanks to the green colour of its waters, Lake Turkana in northern Kenya is also known as the Jade Sea. Caused by algae, the colour shows to particular advantage on the lake's southern shore, where a lava stream from a nearby volcano flowed deep into the basin of the lake before it solidified. Lake water penetrated the brittle igneous rock and permeated through to the depressions in the solidified lava flow, creating a landscape of lakes. Lake Turkana is slightly saline and evaporates quickly in the desert climate of northern Kenya, which is why mineral salts have been deposited along the water's edge.

Lava fountains, Erta Ale
Danakil Desert, Ethiopia

Scientists believe that shortly after earth's formation around 4.5 billion years ago, the planet was covered in an ocean of boiling magma. The lava lake in the crater of Erta Ale gives us an idea of how our planet might have looked at that time. The lava bubbles to a greater or lesser extent, depending on the amount of gas that reaches the surface of the molten mass. Often several lava fountains shoot out through the glowing cracks at the same time.

Lava lakes are very rare. Apart from Erta Ale, Mount Erebus in Antarctica is the only other volcano with one in its crater. A famous lava lake once boiled in the Halemaumau crater on Kilauea in Hawaii. It was active several times over the last 100 years, most recently in 1968, but has since cooled and solidified.

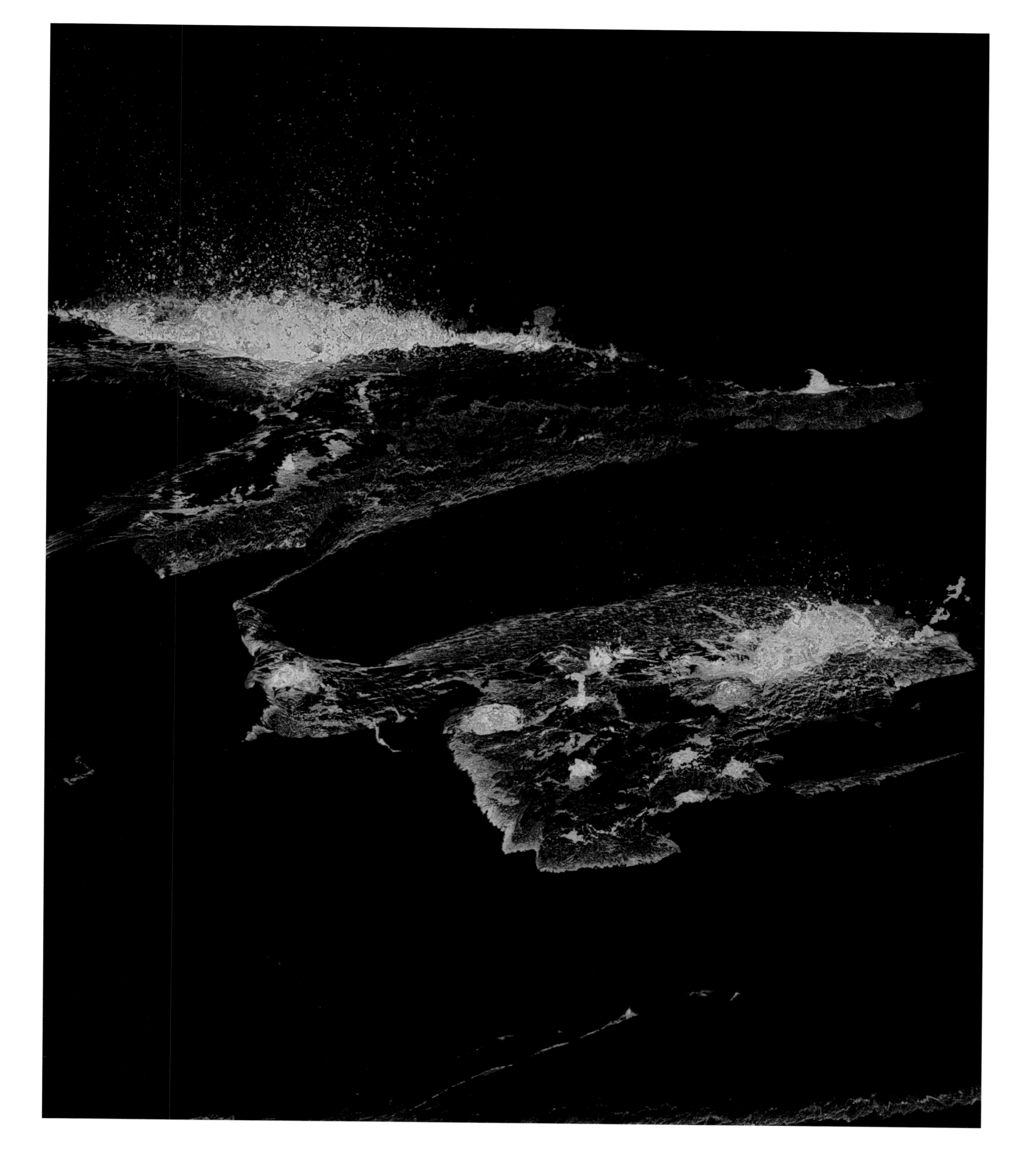

Salt structures of Lac Abbé
Djibouti

Streaks of salt cover the surface of Lac Abbé at the foot of Dama Ali, a volcano on the border between Djibouti and Ethiopia. This lake lies in the centre of the Afar Depression, and is the last in a chain of salt lakes into which the Ethiopian Awash River flows.

Lac Abbé is salty not only because of its high rate of evaporation, but also due to the hot mineral springs that bubble up on its shores. The sediments they leave behind build metre-high stalagmite-like towers that line up along cracks in the ground.

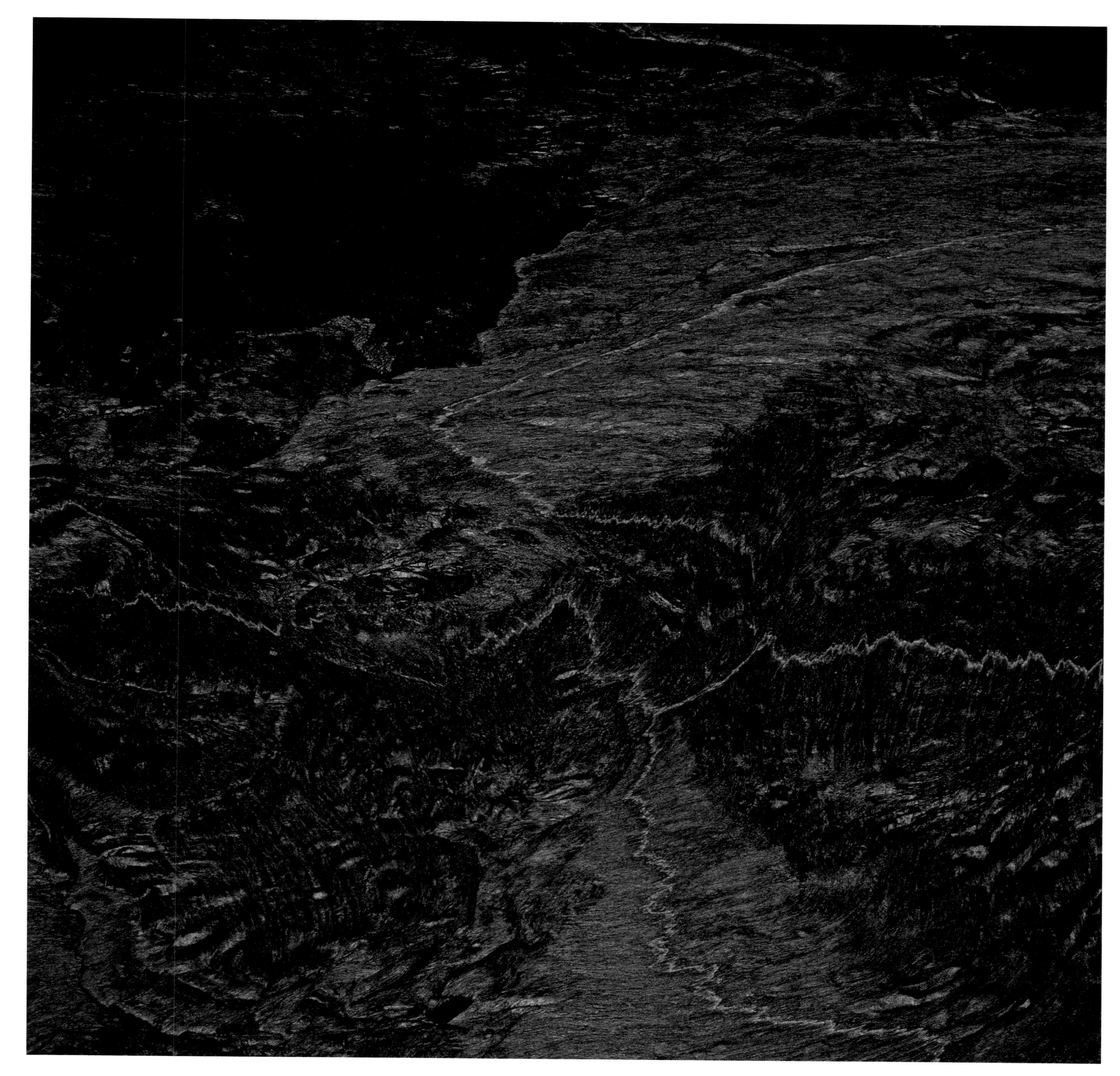

Erta Ale
Danakil Desert, Ethiopia

The skin covering the glowing, molten mass of the lava lake in the crater of Erta Ale volcano is reminiscent of shiny folds of metallic cloth. The lava of this volcano is similar in its chemical composition to the molten material that pours out of cracks in the mid-ocean ridges. For this reason, Erta Ale and the lava deserts that surround it are highly prized by researchers, as they are able to carefully study on dry land what would otherwise require diving boats and a trip to the depths of the ocean to investigate.

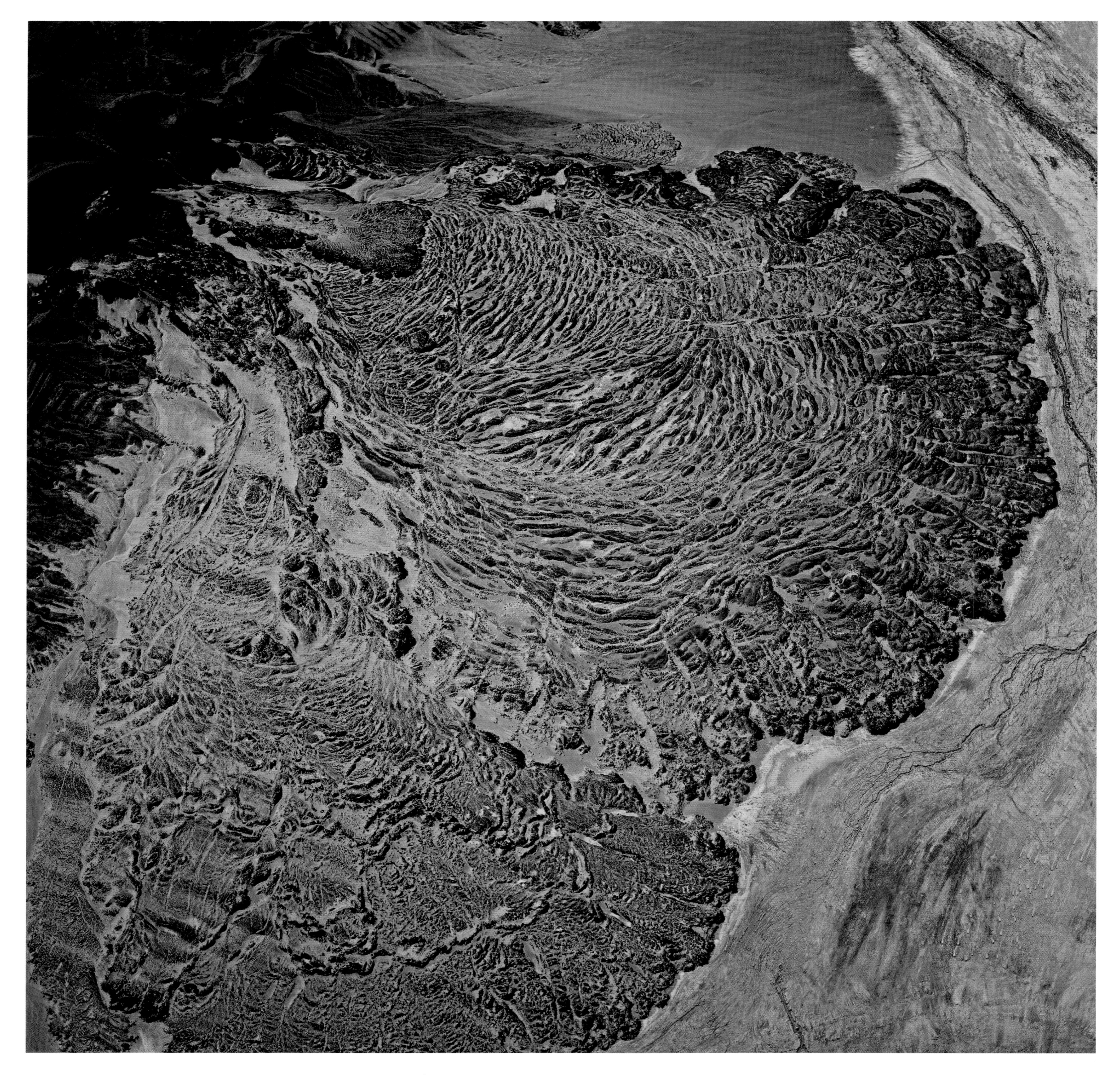

Lava field in the Sugata Valley
Kenya

Lava flows cover the edges of the Sugata Valley in northern Kenya like giant pancakes. The molten rock was thin and fluid, and solidified in rope-like bands, one next to the other. The wind has then blown light-coloured sand onto the black rock, where it has settled, tracing the routes and shapes of the old lava flow.

The Sugata Valley (also called the Sugutu Valley) lies on the central axis of the eastern part of the East African Rift and is separated from Lake Turkana by the Barrier volcanic complex. It is one of the hottest places on earth – midday temperatures here can climb to well over 50 degrees Celsius. Until 3,000 years ago the valley was filled with a lake, which has now almost completely dried up.

Crater in the Sugata Valley
Kenya

Not far from the pancake-like lava fields of the Sugata Valley (opposite) lies this starkly eroded crater. It stands on a thick sheet of dark volcanic rock that has broken up into several long, tilted plates. Such structures are typical of the East African Rift. As in the Afar Depression some 1,200 kilometres to the north, they point to the fact that the earth's crust is constantly being stretched here, allowing molten rock to rise up from the earth's interior and break through the surface. This spreading process is, however, currently considerably more active in the Afar Depression than in the Sugata Valley.

Salt crust, Lake Magadi
Kenya

Crusts and strands of soda minerals float on the water of Lake Magadi in southern Kenya. The lake's basin lies on top of a rift zone in the eastern axis of the East African Rift. The hot springs that feed the lake have a temperature of up to 85 degrees Celsius and carry trona, a salt mineral, out of the bedrock, which crystallizes as the water gradually evaporates under the scorching East African sun. In the dry season, 80 per cent of the surface of the lake, which is 30 kilometres long and three kilometres wide, is covered in a thick, white salt crust.

Salt-loving microorganisms colour the water red over large areas and attract flocks of flamingos, who feed on them.

Salt crust, Lake Natron
Tanzania

Plates of soda salts also float on Lake Natron in northern Tanzania. Like Lake Magadi, which lies 25 kilometres to the north, this lake gets its salt from the bedrock of the East African Rift. This comes partly from the hot springs that bubble up from the floor of the lake and is partly washed in from the rock in the surrounding area during the rainy season. Ol Doinyo Lengai volcano, whose lava contains trona compounds as well as other minerals, rises up nearby.

Lake Natron is 55 kilometres long and 20 kilometres wide. Its depth varies, but reaches a maximum of three metres, making it three times as deep as Lake Magadi, which has a depth of just one metre in the rainy season.

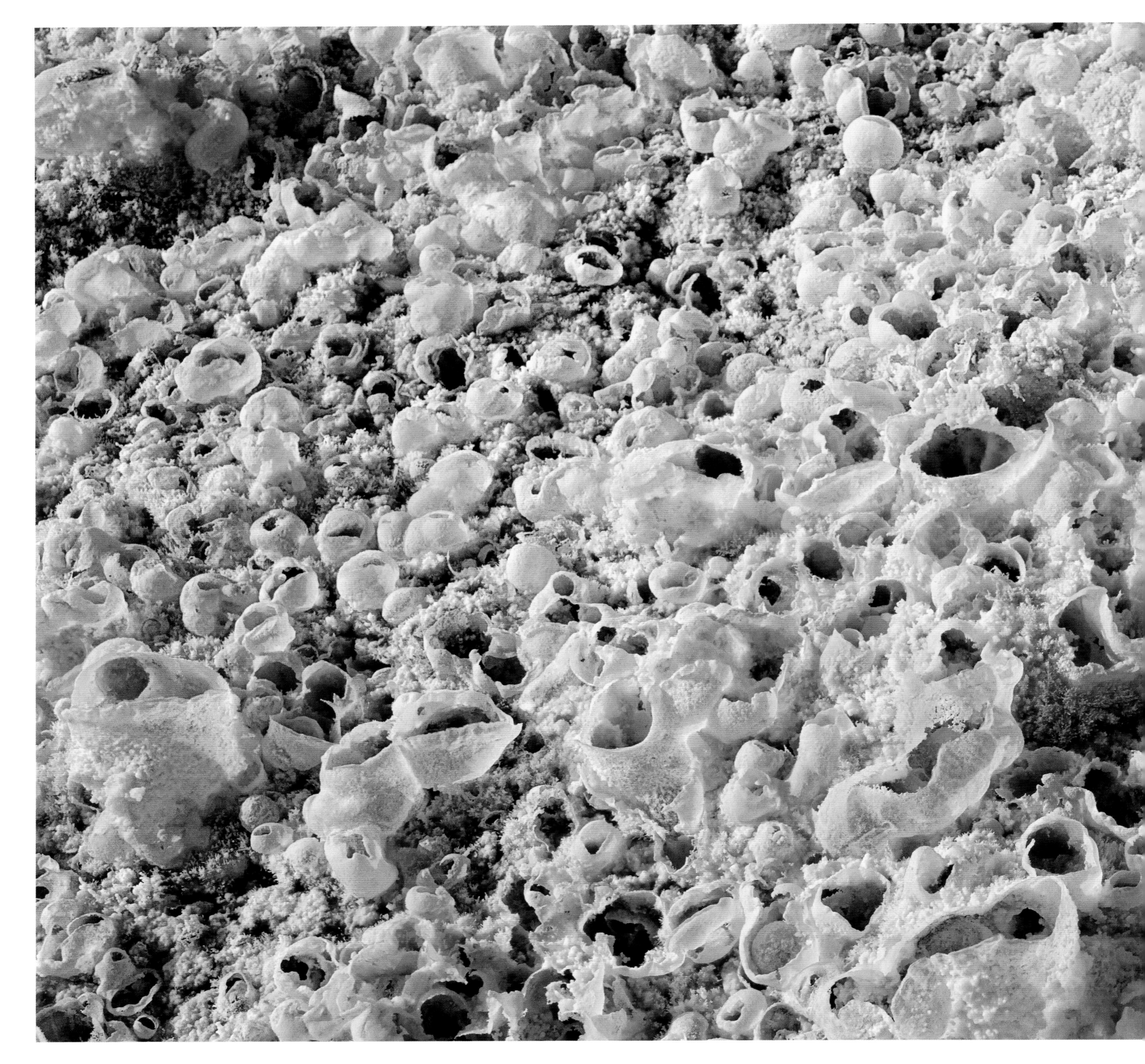

Exhalation structures
Dallol, Ethiopia

These delicate bubbles are created by hot, corrosive gases that emerge from the salty ground of the Dallol geothermal area in northern Ethiopia. The gas brings various minerals up out of the bedrock, which cool and crystallize as soon as they reach the surface. Sometimes very delicate, dome-like structures build up around the exit holes of the gases. Up to eight centimetres high, they look like broken eggshells and a light wind is enough to destroy them. They owe their yellow colour to sulphur and iron minerals.

Nyamulagira
Virunga Mountains, Democratic Republic of Congo

This black landscape on the flanks of Nyamulagira volcano in Congo is only around one year old. The lava that shot out of various craters in fountains and gushed out of hidden fissures in glowing streams left behind a great, crusty sheet of solidified rock. The dense, ancient forest in the surrounding area was burnt to the ground, and only charred tree trunks remained. Fumaroles deposited yellow sulphur on the edge of a crater that was newly created during the eruption, but then just as quickly destroyed by it.

Nyamulagira is a shield volcano, and belongs to the chain of eight Virunga Mountain volcanoes that lie in the border area of Rwanda, Uganda and the Democratic Republic of Congo, famous for its mountain gorilla population.

Kilauea, Big Island, Hawaii
Cascades of glowing lava pour out on the coast of Hawaii's Big Island in the Pacific. Corrosive steam rises up, which the island's residents call 'vog', or volcanic smog.

One of the most striking formations on the planet is the Pacific Ring of Fire, a 40,000-kilometre long chain of mostly highly explosive volcanoes that surrounds the ocean in a horseshoe shape. This is where the Pacific Plate, the largest on earth, dives under its neighbouring plates.

The northern third of this horseshoe consists of the boundary between the Pacific and North American Plates, which runs from the Kamchatka Peninsula in Siberia, with its 160 volcanoes, through the hundreds of volcanic Aleutian Islands and the largely glaciated volcanoes of Alaska all the way down to California. A particularly imposing volcanic mountain range rises up where the small Juan de Fuca Plate is squeezed between the two larger plates. It is being pushed under the North American Plate at a speed of five centimetres a year, where it melts, and the magma created is forced up through the volcanoes of the Cascade Mountains. These include Mount St Helens, whose summit exploded in 1980 during one of the worst volcanic eruptions of the twentieth century.

There are also active volcanoes in the middle of these plates, far from the edges. The Pacific Plate, for instance, drifts at a rate of seven centimetres a year over a hot spot hidden deep in the earth's interior which created the islands of Hawaii, one after the other. Relics of long-extinct volcanoes can be found in the middle of continental North America, such as Ship Rock, which rises over the New Mexico desert. Around 630,000 years ago, a supervolcano erupted in what is now Wyoming. Yellowstone National Park, which contains the largest geyser field in the world, lies in its crater.

North America and the Pacific

Mutnovsky
Kamchatka Peninsula, Russia

Corrosive gas clouds hiss out of the crater floor of the Mutnovsky volcano. They contain a large amount of sulphur, which is deposited around the exit holes, creating a brittle, bright-yellow crust. Sulphur formations shaped like chimneys or columns rise up to a metre in height around some of these fumaroles.

The fumarole fields of Mutnovsky should be approached with great caution. The corrosive gases irritate the eyes and nose, and in some places the ground is not only hot, but also crumbly. Anyone breaking through the crust would suffer serious burns.

Mutnovsky
Kamchatka Pensinsula, Russia

The 2,322-metre high Mutnovsky volcano is a complex structure. It consists of four stratovolcanoes nested into one another on the body of an old shield volcano. The route to Aktivnaya Voronka, the active crater (on the right in the photograph), runs through a valley into which the meltwater of Mutnovsky's glaciers runs off. The glacial stream crosses fumarole fields, causing the water to boil in places.

Mutnovsky is one of the most active volcanoes on Russia's Kamchatka Peninsula. The sulphuric steam that emerges from its active craters reaches temperatures of up to 700 degrees Celsius.

Avachinsky and Koryaksky
Kamchatka Peninsula, Russia

The crater of the 2,751-metre high Avachinsky volcano filled to overflowing during an eruption in January 1991. Two years later, when this photograph was taken, hot sulphuric vapours were still rising from the lava crust.

The flanks of its distant, less active neighbour, the 3,456-metre high glaciated Koryaksky, are marked with distinctive ridges. These barrancos are typical of stratovolcanoes that have not erupted for a long time, allowing rainwater to carve erosion channels out of their steep slopes.

Both volcanoes are only about 25 kilometres from Petropavlovsk, the capital of Kamchatka. They can be dangerous to the city's population when snow and ice melts during eruptions, causing mudslides.

Maly Semiachik
Kamchatka Peninsula, Russia

Filled with a turquoise lake, the Troitsky crater is one of six craters on the long ridge of Maly Semiachik. Sulphuric vapours not only rise up out of the crater walls, but also out of the lake bed, turning the water into a sulphuric acid.

This acid lake was studied in detail in the Soviet era, when there were two volcanological institutes on Kamchatka employing more than 400 scientists. Since then scientists from around the world have continued to observe the lake. It is 140 metres deep, with a constant water temperature of around 40 degrees Celsius and a pH level of 0.4. Like all Kamchatka's volcanoes, which form a 500-kilometre long chain that runs parallel to the peninsula's coastline, the 1,560-metre high Maly Semiachik is part of the Pacific Ring of Fire.

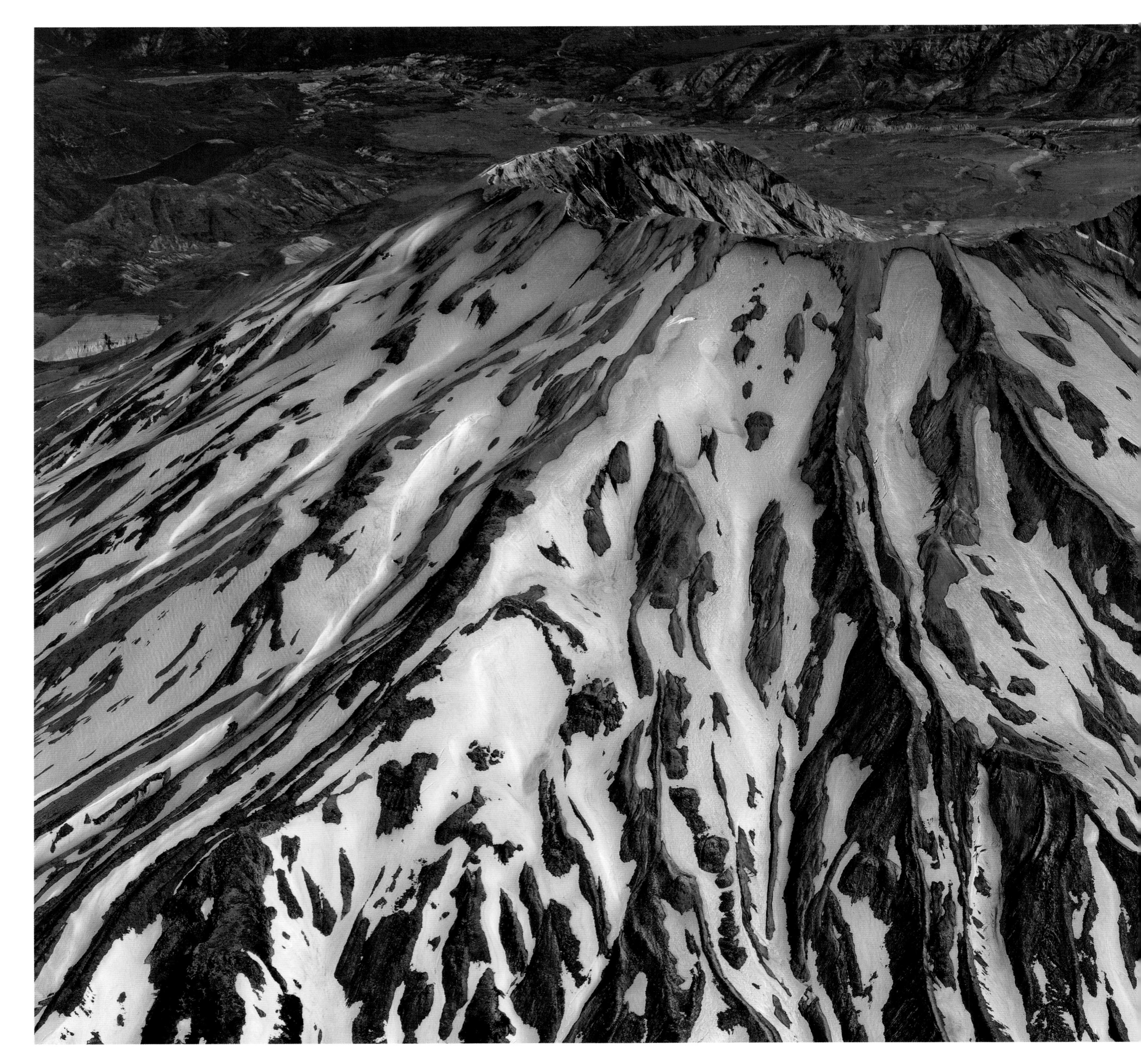

Mount St Helens
Cascade Range, Washington, USA

Thirteen glaciers once covered Mount St Helens, but these disappeared when the volcano exploded on 18 May 1980, losing 400 metres in height. Today the mountain looms like a giant ruin above the Toutle River valley, into which floods of meltwater poured during the eruption, along with red-hot pyroclastic flows.

The devastating outbreak of Mount St Helens is one of the most famous volcanic eruptions of the twentieth century. It was the first time an eruption of this magnitude was captured on camera and broadcast on TV screens around the world. Volcanologists were able to observe what they had long suspected – during their highly explosive eruptions, stratovolcanoes can blast off their summits, with parts of their flanks collapsing as a result.

Yellowstone Canyon
Yellowstone National Park, Wyoming, USA

The colourful walls of Yellowstone Canyon are made up of lava and tuff deposited after the 'Big Bang' 640,000 years ago, when the Yellowstone supervolcano exploded in the North American continent's largest ever eruption. Its collapse left behind an 80-kilometre long and 50-kilometre wide caldera which was gradually filled in by considerably smaller eruptions.

Hot steam and corrosive gases have attacked the once solid volcanic rock, making it brittle. Iron minerals created as a result of chemical reactions have turned the rock orange, red, brown and above all yellow. The Minnetaree tribe's name for the river that cuts up to 300 metres into the layers of lava and tuff was 'Yellow Stone'. The USA's first National Park was later named after this river.

Ship Rock
New Mexico, USA

Ship Rock, a massive rock formation, rises up almost 500 metres over a sandy plain in the middle of the New Mexico desert. It stands at the centre of several bands of dark stone that extend like the spokes of a wheel before disapearing into the desert floor after a few kilometres. The Navajo referred to this distinctive structure as Sa-bit-Tai-e, the winged rock.

Ship Rock and its offshoots are the very last remains of a once 1,000-metre high volcano that was spitting fire around 27 million years ago and gradually weathered and eroded after it became extinct. All that is left is cooled magma that blocked up the vent and fissures of the volcanic cone and solidified where it now stands.

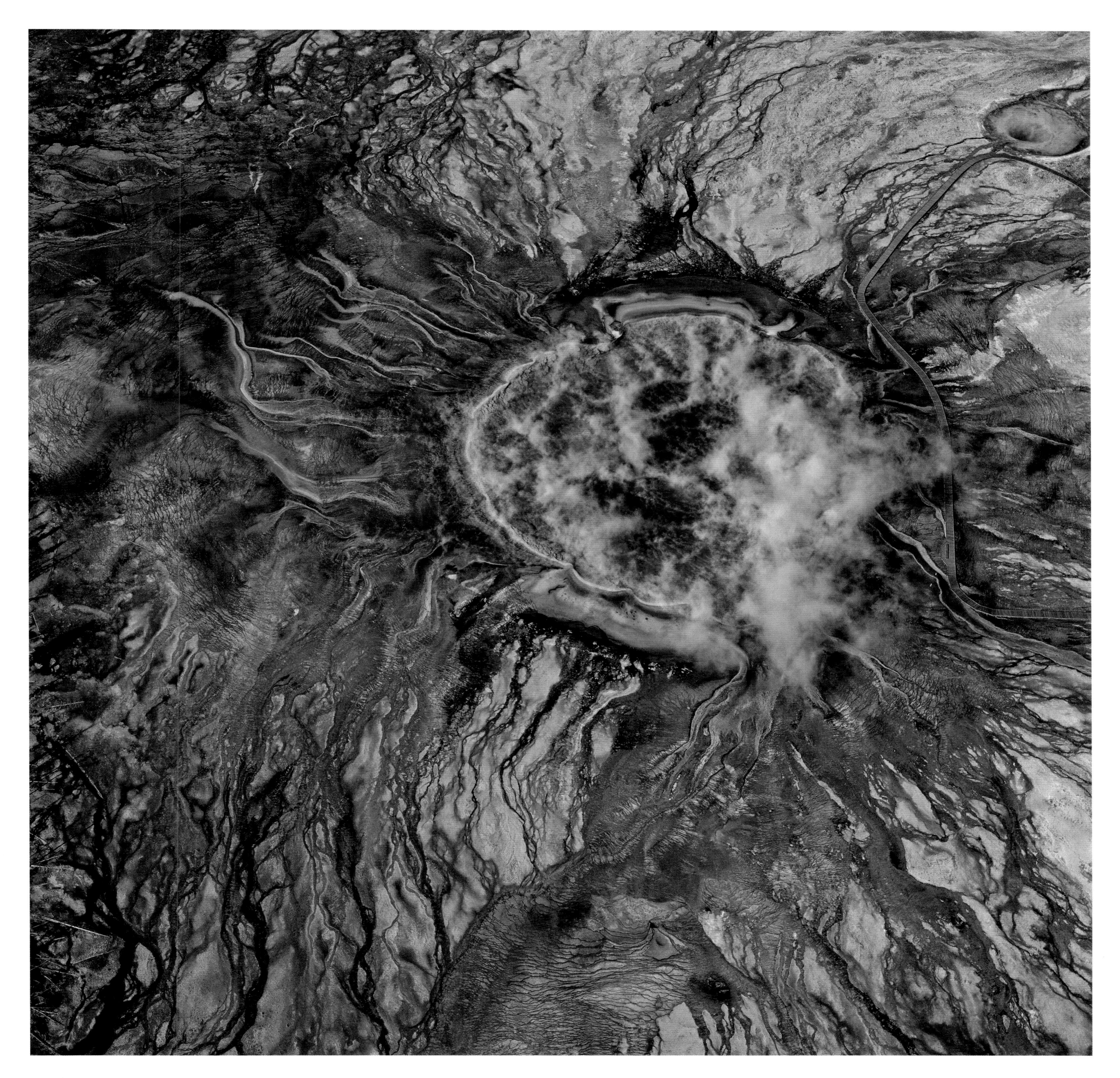

Grand Prismatic Spring
Yellowstone National Park, Wyoming, USA

The water of the Grand Prismatic Spring, the most spectacular of Yellowstone's hot springs, has a temperature of 80 degrees Celsius. Its basin has a diameter of around 80 metres and is about 40 metres deep. The water, which flows out flat around the spring, deposits geyserite. Mats of heat-loving bacteria and algae, which hold carotene, also colour the sinter crust orange and brown.

The existence of the Grand Prismatic Spring was made known by a trapper who was staying in the area in 1840. He could not explain the striking phenomenon, noting in his journal that there was either something special in the air or something chemical in the water. Today it is clear that the hot springs in Yellowstone National Park are signs of volcanic heat in the bedrock.

Toutle River valley
Cascade Range, Washington, USA

An immense quantity of hot material poured into the Toutle River valley at the foot of Mount St Helens during its devastating eruption in May 1980: rubble from the landslide that preceded the eruption, hot ash and lumps of rock from pyroclastic flows and mud from lahars that rushed down the volcano's flanks. In places the valley was filled to a height of 200 metres. Hot gases trapped in the rubble exploded and blew open craters in the freshly deposited material. Later, channels were cut into the rubble by ground- and rainwater, which collected in depressions with no drainage, creating lakes. It is very difficult for plants to get a foothold in this kind of chaotic landscape, as the loose material holds barely any water and is easily washed away during rainfall.

Morning Glory
Yellowstone National Park, Wyoming, USA

The pool of the Morning Glory hot spring opens up like a blossoming flower. Along with the Grand Prismatic Spring, it is one of the most well known of the roughly 10,000 hot springs in Yellowstone National Park. The water in the centre of the pool – which has a diameter of around seven metres – has a temperature of almost 100 degrees Celsius. This spring pool owes the colour on its edges to heat-loving bacteria and algae.

The Yellowstone geothermal area is the largest of its kind in the world; larger than similar areas in Iceland, Chile, on the Kamchatka Peninsula in Russia and on the North Island of New Zealand. Apart from hot springs, it also includes boiling mud pools and around 250 active geysers.

Mammoth Hot Springs
Yellowstone National Park, Wyoming, USA

The water that streams out of the ground from Mammoth Hot Springs is very rich in calcium carbonate and, as soon as it comes into contact with cold air, it deposits its load. The calc-sinter, also known as travertine, creates shallow, stepped pools over which the water flows in little cascades.

Sinter terraces can change dramatically over very short periods of time. During long dry spells, the water volume drops and sections of the terraces begin to crumble and discolour. When there is a large supply of water, however, so much calcium carbonate can be deposited that the travertine crust becomes 20 centimetres thicker in places over the course of a single year.

Iliamna
Alaska, USA

Iliamna is part of a chain of glaciated volcanoes that runs along the coast of Alaska and on into the arc of the Aleutian Islands. The volcano's eruptions are not particularly well documented, and the last major eruption probably occurred long before Europeans laid eyes on this 3,053-metre high stratovolcano. It is known for landslides that roll down its icy slopes, shaking the earth to such an extent that tremors are picked up on seismic instruments in Alaska's Volcano Observatory in Anchorage, 230 kilometres away. Iliamna is under close observation, as the fumaroles below its summit indicate that the volcano is active. The sulphuric steam cloud above can get so large that it almost looks as if there had been an eruption.

Kilauea
Big Island, Hawaii, USA

Just two days before this photograph was taken on Hawaii's Big Island, a large section of the coastal area where lava streams from the Kilauea volcano flow into the Pacific Ocean broke off. These 'bench collapses', as Hawaiians call them, occur on a regular basis, as the areas of freshly solidified lava are brittle and unstable. Slabs the size of football pitches can pull away and sink into the sea, leaving behind steep walls such as this one, where a lava tunnel – one of the underground outflow pipes – has been cut open. The glowing, molten material that shoots out quickly solidifies to create a landscape that looks as if made of frosting. Usually hidden by clouds of steam, it is only rarely seen.

Kilauea
Big Island, Hawaii, USA

Lava has been continously spewing out of Puu Oo, a flue on the flanks of the Kilauea volcano, since 1983. In 1987 the glowing streams reached the coast, 11 kilometres from the crater, and have been pouring into the Pacific Ocean ever since, constantly extending the coastline further into the sea. Scientists from the Hawaiian Volcano Observatory have worked out that 3.1 cubic kilometres of lava had flowed into the sea by January 2007 and that, within this 30-year period, the island has grown by around 201 hectares (around two square kilometres) – that is around the size of the Principality of Monaco on France's Mediterranean coast, or 275 football pitches.

Kilauea
Big Island, Hawaii, USA

The lava of Kilauea has a temperature of around 1,200 degrees Celsius and is almost as thin as water. It solidifies in scallop-like shapes or twisted, rope-like strands. Sometimes it also encases itself as it flows over the flat volcanic slopes, because the edges of the lava streams cool down more quickly than the centre. It initially creates a channel, which turns into a tunnel if the surface of the flow also begins to solidify. The freshly solidified lava is very porous in many places, and holes can appear in the roof of the tunnel. These skylights offer a view of the glowing, molten material which is well insulated in these underground tunnels, remaining hot and fluid. The lava is able to cover long distances in this way.

Mount Wrangell
Alaska, USA

A blue meltwater lake glistens in the ice of Mount Wrangell, not far from the glaciated crater at its summit, out of which a few fumaroles rise up. On clear, cold, windless days, the clouds of steam are visible from a great distance. Small ash clouds sometimes erupt out of the crater, blackening the surrounding ice.

Although it lies on a subduction zone, the 4,317-metre high ridge of Mount Wrangell is in fact a gigantic shield volcano, not an explosive one, with a lava volume of around 1,000 cubic metres – that is 31 times more than that of Mount St Helens. Mount Wrangell last erupted in 1930. There were signs of an impending eruption again in 1980, but the volcano quietened down.

Mount Spurr
Alaska, USA

A green acid lake fills the active crater of Mount Spurr, at 3,374 metres the highest glaciated volcano of the Aleutian chain. The fumaroles streaming out of the crater wall keep the water warm enough for it to remain liquid.

The crater opens up on the flanks of Mount Spurr around 1,000 metres below the volcano's summit. During the two most recent eruptions, in 1953 and 1992, powerful eruption clouds shot out of this vent, and climbed over 10 kilometres into the air. Some 130 kilometres away, Anchorage, the largest city in Alaska, was hit by falling ash and, in 1992, the city's airport, the most important in the state, even had to be closed down for several days.

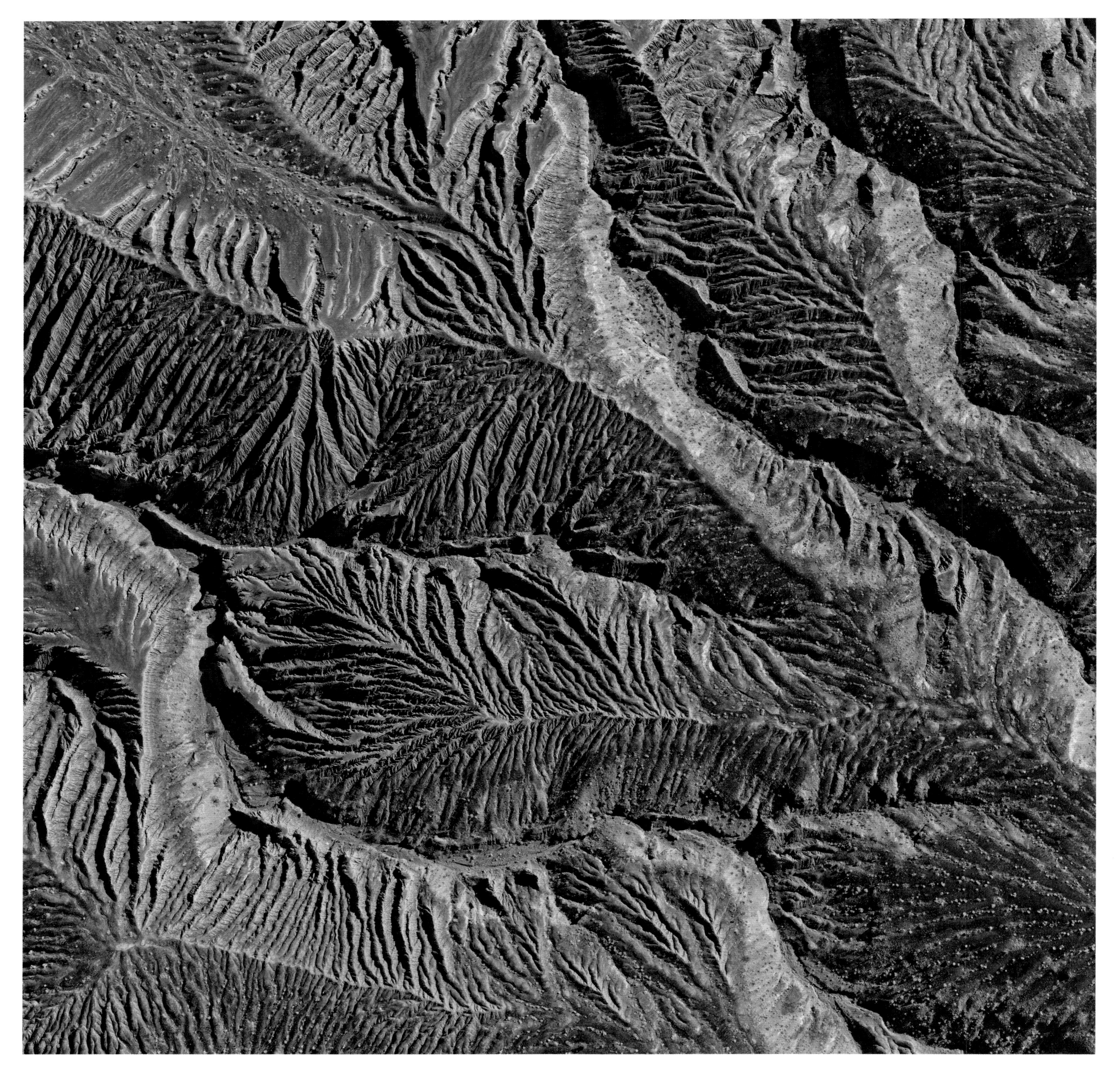

Ubehebe Crater area, Death Valley
California, USA

About 2,000 years ago, large volcanic eruptions took place in Death Valley. As the earth's crust slowly spread, fissures broke open, and magma pushed up. Enormous steam explosions occurred when the magma met groundwater beneath an alluvial fan in the north of Death Valley. The glowing molten material was completely shredded and pulverized as it was catapulted into the air and fell back down to earth in the form of black ash mixed with large chunks of solidified lava.

The explosion left behind a layer of ash up to 50 metres thick over Ubehebe crater and its neighbouring craters, covering an area of 15 square kilometres. The region's rare but heavy rainfalls have since carved out the landscape.

Artist's Palette, Death Valley
California, USA

The colourful layers at the foot of Death Valley's Black Mountains are also volcanic in origin, but they are much older than those in the Ubehebe crater region. They date from the period between two million and 65 million years ago during which the mountains on either side of the valley were created, and now long-since extinct volcanoes erupted. Several rounds of ash rained down here, where the famous Artist's Drive now runs through the colourful, hilly landscape. The colour was created when the volcanic ash was chemically altered, first by hot gases and later as a result of weathering. The red, pink and yellow colours are evidence of different iron minerals, while manganese compounds are violet and the green comes from clay containing mica.

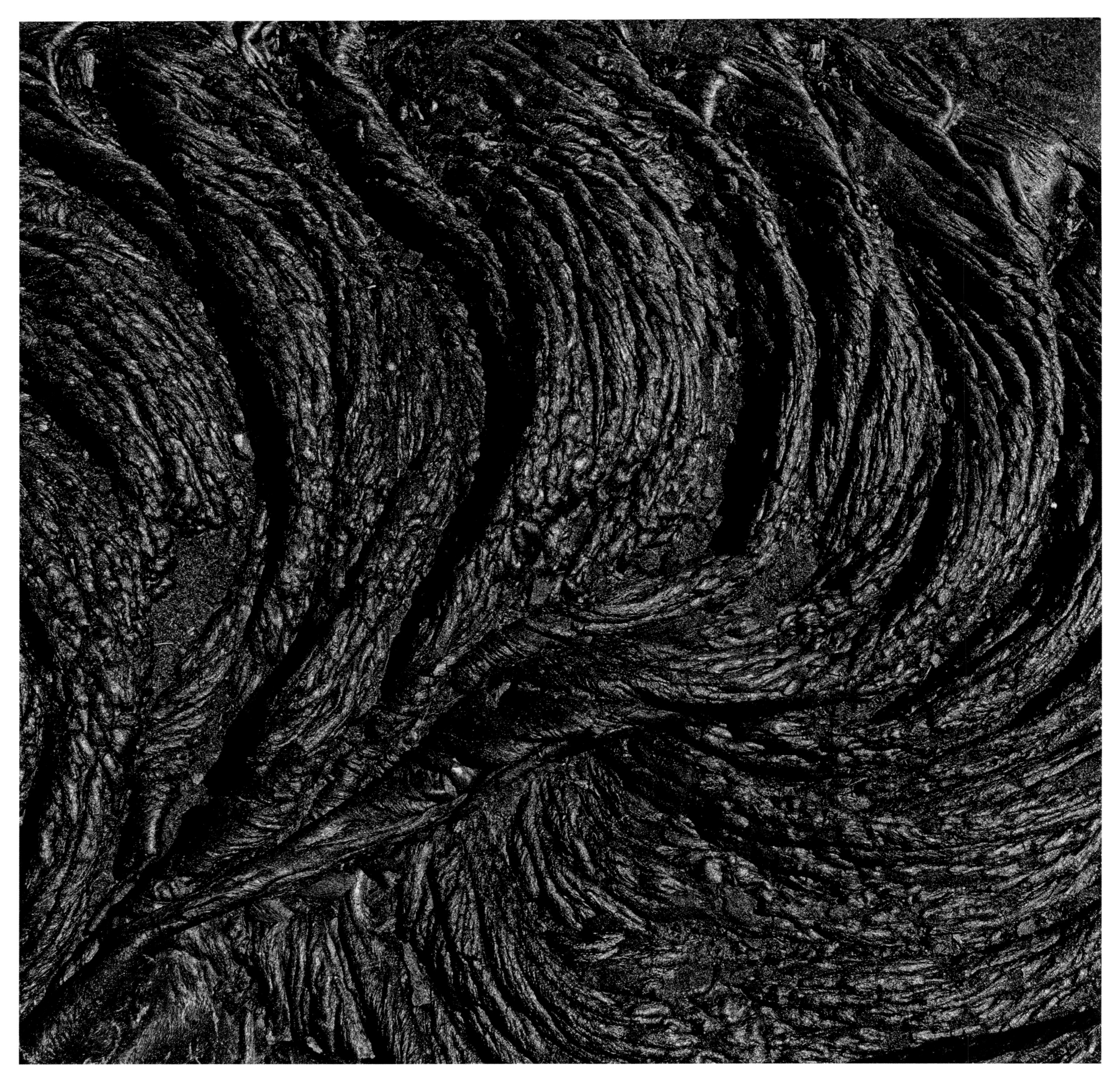

Pahoehoe lava, Kilauea
Big Island, Hawaii, USA

When fluid, runny lava flows solidify, a black skin first forms on top of the glowing molten rock. As the lava continues to flow, this skin is pushed together and layers up in folds. A brittle crust with a rope-like structure is created where the lava flow comes to a standstill and cools. Hawaiians call this type of lava pahoehoe lava, which means 'lava on which one can easily walk barefoot'.
In contrast, the surfaces of solidified flows of viscous lava contain loose, sharp-edged bits of cinder, and Hawaiians call this type aa lava, after the involuntary sound a person makes when stepping barefoot on jagged rock. Both these Hawaiian phrases are now used throughout the world to describe the solidified crust of lava flows.

Puu Oo crater, Kilauea
Big Island, Hawaii, USA

The strange, 300-metre high Puu Oo Crater rises up over lava fields on the flat southeastern flank of the Kilauea volcano. This is where Hawaii's most recent eruption began in 1983, and continues today. The crater was frequently the backdrop for spectacular lava fountains that shot up 500 metres into the air, and the ash and cinder cone of Puu Oo was created in less than three years as a result. The Puu Oo crater sometimes holds a lava lake. The molten mass flows from the crater's vent to the sea via an underground tunnel system.

Puu Oo is a Hawaiian word. *Puu* means hill and *Oo* is the name of an extinct bird whose shining gold feathers the island's rulers used as adornment.

Hot spring
Yellowstone National Park, Wyoming, USA

The hot water in the bedrock of Yellowstone National Park draws out minerals, depositing them again as soon as it comes into contact with cool air. This is how pale, coarse sinter crusts are created. The whitish streaks of algae and bacteria acclimatized to the high water temperature are almost indistinguishable. Alongside their brown relatives, they colonize channels just a few centimetres wide that run through the sinter crust.

The heat in the Yellowstone caldera comes from a hot spot at a depth of 200 kilometres. Magma rises up and gathers around five kilometres below the earth's surface. The temperature 200 metres below the surface is around 200 degrees Celsius, 30 times hotter than in non-volcanic areas.

Painted Hills
John Day Fossil Beds National Monument, Oregon, USA

There have been volcanoes in the Oregon area for 30 million years, blasting huge amounts of ash into the sky. Winds carried the ash to where the John Day Fossil Beds National Monument now lies. This volcanic ash built up, layer after layer, continually burying the marshes and forests that flourished in the moist and warm tropical climate of the period.

The heavy storms that rain down here today carve gullies into the soft layers of ash and, over time, have created the striped landscape of the Painted Hills (above and following pages). The yellow and red layers owe their colour to eroded volcanic materials, while the dark blurry flecks are the remains of dead vegetation.

Great Fountain Geyser
Yellowstone National Park, Wyoming, USA

Every eight to 12 hours, the Great Fountain Geyser in Yellowstone National Park catapults fountains of hot water and steam up to 35 metres into the air, making it one of the biggest and most spectacular geysers in this famous geothermal area. Its central basin, out of which the fountains erupt, has a diameter of over three metres, and the Great Fountain Geyser's eruptions are correspondingly dramatic. Forty minutes before an eruption, steaming water starts to spill over the edges of the basin. Suddenly the geyser shoots out enormous amounts of water. Pulsing jets of water seem to try and outdo each other, while loud whooshing and hissing noises fill the air. After about an hour, the fountains become weaker before finally drying out.

Wizard Island, Crater Lake
Cascade Range, Oregon, USA

The volcanic explosion that took place where Crater Lake now lies, 7,700 years ago, was 40 times more powerful than that of Mount St Helens in 1980. Mount Mazama, a 4,000-metre high volcano, collapsed and left behind a 1,000-metre deep caldera with a diameter of almost 10 kilometres. While small eruptions continued to occur on the caldera's floor, the basin filled with rain and meltwater from winter snowfall, and Crater Lake and Wizard Island were formed. The small volcanic cone rises around 230 metres above the surface of the water.

The last eruptions on the lake bed ended 4,800 years ago, and the volcano has been dormant since. With a depth of 590 metres, Crater Lake is the deepest lake in the USA. Its water is very clean and clear, making it an intense blue.

Wheeler Canyon
Crater Lake area, Cascade Range, Oregon, USA

Spiky stone columns line Wheeler Canyon, which lies around eight kilometres southeast of Crater Lake. The canyon cuts through the 200- to 300-metre deep layers of ash and pumice that were thrown out of the Mazama volcano during its devastating eruption 7,700 years ago. This material was still hot long after being deposited on the flanks of the volcano. Trapped corrosive gases escaped and streamed out of the hot tuff as fumaroles. This process cemented the loose ash and pumice stone along the outlet channels, making the tuff there more resistant to weathering and erosion. The exit holes of the fumaroles are still visible at the top of many pinnacles. These stone columns can be up to 60 metres high.

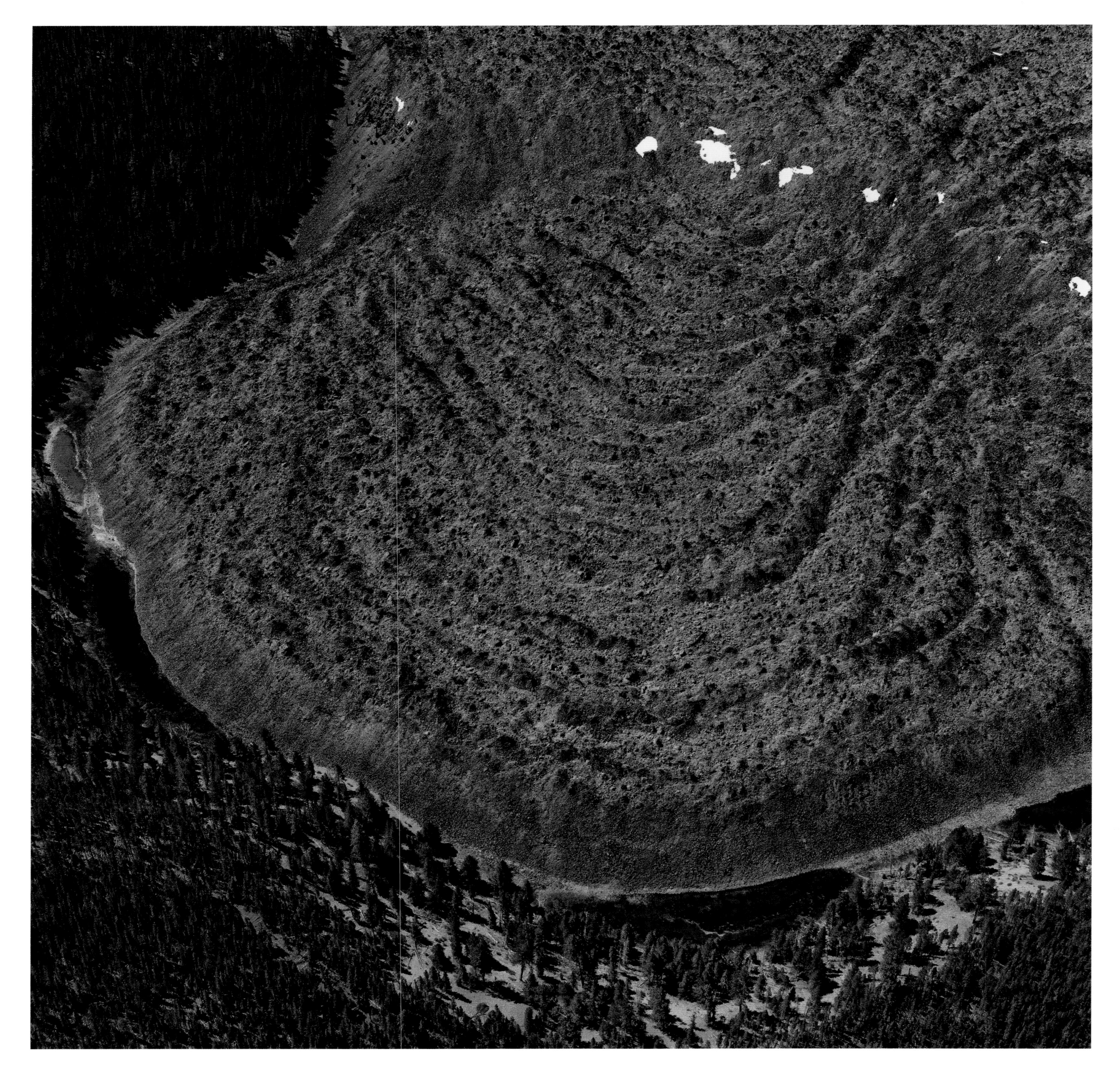

Big Obsidian Flow
Newberry volcano, Oregon, USA

This lava flow consisted of gas-poor, silica-rich lava that flowed out of one of the Newberry volcano's many craters around 1,300 years ago and cooled very quickly. The minerals in the molten rock did not form a crystal lattice, but solidified into an amorphous mass of black glass. The lava flow covers an area of more than 2.5 square kilometres and is more than 20 metres thick in places.

The eruption that led to the Big Obsidian Flow was the last in this area. Since then the giant shield volcano, which has a diameter of 32 kilometres, has been dormant. There are 400 cinder cones on its back. Scientists believe that there is magma just five kilometres below the earth's surface here.

Hot spring
Yellowstone National Park, Wyoming, USA

Different types of heat-loving algae and bacteria populate the sinter crust that covers the ground around the hot springs in Yellowstone National Park. They form gelatinous carpets and streaks up to half a metre long, either white, green or black, that move around in the water currents.

Biotechnology companies from around the world search the hot springs of Yellowstone for microorganisms that can be commercially exploited. More than 10 years ago, the bacteria *thermas aquaticus* was discovered there, whose enzymes are used today in genetic engineering laboratories for the duplication of genetic material.

Na Pali coast
Kauai, Hawaii, USA

Until 520,000 years ago, Kauai, the oldest of the Hawaiian islands, was a gigantic active shield volcano. The volcano became extinct when the Pacific Plate moved northwest, so that the connection to the hot spot in the lithosphere that served as its source of magma was lost.

Since then Kauai has been exposed to the forces of weathering and erosion, which have had a particularly marked effect on the remote Na Pali coast in the north of the island. Rains driven in from the Pacific pour down here, and the water has carved deep channels and sharp-ridged cliffs up to 1,200 metres high out of the volcano's flanks. The rolling ocean waves also constantly erode the front of the bluffs.

Fumaroles, Poás, Costa Rica
The crater walls of active volcanoes like Poás are extremely brittle. They are not only fractured during earthquakes that accompany eruptions, but also affected by hot, corrosive sulphuric gases.

Between North and South America, there is a 'crumple zone' where five lithospheric plates meet. The Caribbean Plate lies at the centre of this conglomeration, pushed and shoved from all directions. The North American Plate rubs against it in the north, the South American and Nazca Plates in the south. From the east, a piece of the Atlantic Ocean floor creeps under the Caribbean Plate at a speed of around 2.5 centimetres a year, while the small Cocos Plate dives underneath it from the Pacific, in the west. It pushes itself underneath the Caribbean Plate at a speed of nine centimetres a year, making it one of the fastest diving plates on earth. Higher speeds can only be found further south, along the coasts of Ecuador, Peru and Chile, where the Nazca Plate dives under the South American Plate at a speed of 10 to 11 centimetres a year. Some of the highest active volcanoes in the world rise up here, including the 5,897-metre high Cotopaxi in Ecuador.

The eruptions of almost all the volcanoes in this region are highly explosive. Up to 90 per cent of the molten rock pushed up to the earth's surface through their vents is either catapulted high into the air or hurtles down the sides of the volcanoes in pyroclastic flows. Only around 10 per cent of the magma pours out in the form of lava.

South and Central America and the Caribbean

Tungurahua
Cordillera Central, Ecuador

Tungurahua soars 5,023 metres up into the sky. It belongs to the chain of 20 fire-spewing mountains that line Ecuador's famous Avenue of the Volcanoes. Its black, smoking crater can frequently be seen above the clouds blown in from the nearby Amazon basin.

This stratovolcano is 14,000 years old and is known for intense eruptions accompanied by massive ash clouds as well as lava avalanches and – if the ice on the mountain melts – streaming mud that destroys the fields on its slopes and endangers the villages and towns at its foot. In November 1999, 22,000 people had to be evacuated from the area.

Arenal
Costa Rica

Arenal is the youngest and most active stratovolcano in Costa Rica. Ash clouds have billowed out of the higher of its two craters and fiery rocks rumbled down its flanks since the start of its last eruption phase in 1968. The cone has grown by around 200 metres since then. Like all volcanoes in the chain along Central America's Pacific coast, this 1,680-metre high giant is the result of subduction. The Cocos Plate pushes under the Caribbean Plate at a speed of nine centimetres a year here. It melts at a depth of around 100 kilometres, causing magma to rise and spew out of volcanoes from Mexico to Costa Rica. Arenal first erupted around 7,000 years ago. It throws giant rocks, so-called bombs, out of its crater with such force that they hit the ground several kilometres away.

Rincón de la Vieja
Costa Rica

Like Arenal, Rincón de la Vieja is one of the most closely watched volcanoes in Costa Rica, particularly its active crater. Filled with an acid lake, it has a diameter of around 500 metres and lies at a height of 1,806 metres. Hot, corrosive mudslides rush down the mountain during eruptions, endangering villages and plantations at its foot. The last catastrophe of this type occurred in 1998.

The active crater is just a small part of the large Rincon de la Vieja volcanic complex, which consists of nine eruption centres. It includes the large crater bowl of the West Crater (following pages), which is covered in thick, tropical forest. The rainwater that collects here has broken through the crater wall and cascades down the flanks of the old, long-since dormant volcano into the valley.

Poás
Costa Rica

The active crater of Poás is 1,300 metres wide and 300 metres deep. It contains a lake, Laguna Caliente, whose water has a temperature of between 40 and 50 degrees Celsius. Fumaroles make Laguna Caliente one of the most acidic lakes on earth, with a pH level approaching zero.

Poás is a 2,708-metre high stratovolcano. The corrosive steam clouds that constantly stream out of its crater, and are blown over the edge by the prevailing west wind, cause acid rain to fall onto the east flank of the mountain, damaging vegetation across a 10-kilometre wide area.

Poás
Costa Rica

A lava dome stands on the shore of Laguna Caliente, the acid lake in the middle of the Poás volcano. The fumaroles here, whose sulphur is deposited on the hot rock, reach temperatures between 300 and 800 degrees Celsius.

The most recent eruption occurred here in January 2008. From a safe distance a park ranger was able to observe how a 200-metre high fountain of mud and steam suddenly shot up from the middle of the lake and then fell back in, whereupon the water changed colour, turning from turquoise to milky white.

Pacaya
Guatemala

The MacKenney Cone, the steaming main crater of the 2,552-metre high Pacaya volcano, casts an impressive shadow over its surroundings in the clear morning light. It has been constantly active since 1965. Small streams of lava flow out of cracks in the flanks of the crater cone, gradually filling up the gorge between Pacaya and its neighbouring crater, Cerro Chino, which last erupted in the nineteenth century.

Pacaya is one of Guatemala's most active volcanoes. It is easy to climb and has therefore become one of the country's tourist attractions.

Pacaya
Guatemala

Sulphur glows yellow and orange-red in the MacKenney Cone, the most active of Pacaya's three craters. A thick gas cloud constantly billows from its mouth. The crater occasionally shoots fountains of glowing shreds of lava hundreds of metres up into the air, which cool as they fall to earth and then roll down the crater's steep flanks. This loose material is stabilized by lava flows that sometimes stream straight from the crater, but more often emerge from cracks in the flanks of the cone.

Mud pool
Rincón de la Vieja, Costa Rica

Hot springs and mud pools bubble up at the foot of the Rincón de la Vieja volcano. They all lie along the volcano's arid Pacific side and therefore often dry out. Sulphuric mineral crusts cover the hot ground and larger blocks of volcanic rock containing iron start to rust.

The name Rincón de la Vieja, which means 'the old woman's lair', comes from a local legend. An old woman is said to be hiding among the volcanic craters. Afraid of her tiger's teeth and fiery eyes, the people among whom she lived had cursed her as a witch and chased her away. She fled to the volcano, where she still causes mischief to this day.

Aa lava, Pacaya
Guatemala

The lava that pours out of the flanks of the Pacaya volcano is often very viscous. Scientists call this block lava or aa lava – now in use worldwide, the Hawaiian name for this type of lava comes from the exclamation of pain a person walking over it barefoot would make. The name fits: the surface of such flows is strewn with loose cinder pieces that are both porous and sharp-edged, and break very easily when stepped on.

The temperature of lava can be estimated by its colour. If it is glistening yellow, it will be over 1,200 degrees Celsius, while glowing red areas have a temperature of around 1,000 degrees Celsius. Once it drops below 900 degrees Celsius, it turns black.

Cotopaxi
Cordillera Central, Ecuador

At 5,897 metres, Cotopaxi is one of the highest volcanoes on earth. Its name comes from the native Quichua language and means 'smooth nape of the moon'. Since 1758 it has erupted around 50 times, most recently in 1904. While occasional earthquakes have been registered in its vicinity since then, the mountain has always quietened down again.

Cotopaxi's eruptions are dangerous, because the glowing ash and lava causes the volcano's ice cap to melt. Mudslides pour down over its slopes and far into the surrounding area, ripping through everything in their paths. Mudslides caused by the eruption in 1877 had a reach of 100 kilometres.

Guagua Pichincha
Cordillera Central, Ecuador

The crater of Guagua Pichincha volcano is six kilometres wide. The hill in its centre is an old lava dome that has blocked the stratovolcano's vent since its last devastating eruption in 1660. Subsequent, much smaller, explosions formed the small crater. Fumaroles are a clear indication that Guagua Pichincha is as active as it ever was.

Only a chain of hills separates the 4,784-metre high volcano from Quito, the capital of Ecuador. Although the volcano is closely monitored to protect the population, two volcanologists who were taking gas samples from fumaroles died when the volcano suddenly erupted in March 2000. Light earthquakes had preceded the eruption, but the scientists could not be warned in time.

Botos crater, Poás
Costa Rica

Just 800 metres away from the bare, active crater of the Poás volcano, a second crater rises up, almost hidden by thick cloud forest. This crater, Botos, has a diameter of around 400 metres and is filled with cold, crystal-clear rainwater.
According to rock studies, Botos was created just 8,300 years ago and last erupted 7,500 years ago.

Santiaguito
Guatemala

Santiaguito's cone is only around 250 metres high and it is not really an independent volcano. Nevertheless, it is the most dangerous volcano in Central America. Since it was created in 1922, its eruptions have caused more than 5,000 deaths.

Santiaguito is a lava dome that grew out of the collapsed flanks of the mighty 3,772-metre high Santa María volcano. It consists of a total of four tightly packed domes, one of which, called El Caliente, currently shoots ash clouds hundreds of metres into the sky. In 1929, a part of this hot structure collapsed and fiery streams poured over the surrounding farmed lands towards the Pacific. Three thousand people died during this eruption alone.

Fuego
Guatemala

Fuego is not only the most active volcano in Guatemala, but in all of Central America. It is 3,763 metres high, and forms part of a twin volcano complex with its neighbour, Acatenango, which is 200 metres higher and has been dormant since 1972. Both sit on the body of an old volcano that collapsed during a catastrophic eruption in prehistoric times.

The first recorded eruption of Fuego occurred in 1524, at the beginning of the Spanish colonial period in Guatemala. The volcano has erupted 60 times since, and pyroclastic flows have often poured down its flanks. In 2008, when this photograph was taken, it was only moderately active. Gas clouds hung over its crater and it spat small ash clouds into the air around once every 20 minutes.

Soufrière Hills volcano
Montserrat, West Indies

Resembling steaming piles of rubble, lava domes look harmless but are very dangerous. Hot and unstable, they can collapse and release glowing avalanches that sear everything in their path. Sometimes up to 13 cubic metres of solidified lava, with temperatures of around 850 degrees Celsius, are squeezed out of the vent of the Soufrière Hills volcano every second, creating spiky rock needles and stumpy columns that can grow up to 50 metres high. Freed from the enormous pressure in the mountain's interior, these hot formations can suddenly crumble and collapse.

Dormant for 400 years, the volcano awoke in 1995. Since then a lava dome has grown that keeps breaking apart, producing deadly pyroclastic flows.

Soufrière Hills volcano
Montserrat, West Indies

Only at dusk can you tell that the rocks falling from the lava dome on the 915-metre high Soufrière Hills volcano are red hot. By day, only bouncing grey stone blocks can be seen, stirring up dust every time they hit the ground. In November 2000, when this photograph was taken, a giant stone column was visible in the gas cloud enshrouding the volcano's summit. According to scientists in the Montserrat observatory, it reached a height of 70 metres before it collapsed.

The last major dome collapse occurred in December 2008, releasing a kilometre-high ash cloud and pyroclastic flows. Due to the constant threat of eruptions, Montserrat's 6,000 residents have had to withdraw to the northern third of the 16-kilometre wide and seven-kilometre long tropical island.

Soufrière Hills volcano
Montserrat, West Indies

Most of the pyroclastic flows that peel away from the lava dome of the Soufrière Hills volcano roll down through the valley of the Tar River and pour into the sea, where they create a delta.

The houses on the volcano's flanks have been destroyed and have lain abandoned for many years. The fertile fields cannot be cultivated as they lie in an exclusion zone covering the southern two thirds of the island. The British Government set up a volcanic observatory on the island, a British Overseas Territory, so that scientists can monitor the volcano and ensure the safety of the local population.

Soufrière Hills volcano
Montserrat, West Indies

As the lava dome on the Soufrière Hills volcano grows higher, small pyroclastic flows sometimes thunder down the flanks of the volcano. These avalanches of corrosive gas, hot ash and glowing lumps of rock reach speeds of 200 to 300 kilometres an hour. The rocks shatter, releasing hot gas trapped in their fine pores. This drives the hot dust and ash mixture into the air, and a grey-brown eruption cloud billows up.

Typical of stratovolcanoes, pyroclastic flows are one of the most deadly volcanic phenomena. A dome collapse in June 1997 set off the first large pyroclastic flows on the Soufrière Hills volcano. Although the population was warned, not everyone managed to escape. Twenty-three people lost their lives.

Volcanology

Active, dormant or extinct

Many volcanoes on earth are considered active even though they are not currently spewing any lava or ash – they are dormant or sleeping. Sometimes just a few months or years will pass between eruptions; but it can also be a matter of decades, centuries or even of many thousands of years. Most of the 115 active volcanoes in the East African Rift do not exhibit their might as clearly as the fiery Erta Ale, Mount Nyiragongo or Ol Doinyo Lengai. Yet they still count as active; even though it is impossible to tell that magma is simmering within and beneath them. The craters on the islands in Lake Turkana in northern Kenya can be included in this category, as can many volcanoes in Ethiopia, and even Mount Kilimanjaro, the highest volcano in Africa. No one knows exactly when lava last flowed from Kilimanjaro, whose activity for hundreds of years now has been reduced to a few fumaroles – gushings of sulphurous gases. Nevertheless, the mountain still does not qualify as extinct.

Scientists from the Smithsonian Institution in the United States, who are documenting eruptions worldwide as part of their Global Volcanism Program, have established statistical criteria in order to grade all volcanoes as active or extinct. Depending on which of their criteria one uses, there are either 550 or over 1,500 active volcanoes on earth. The first figure is derived from counting those volcanoes that have erupted at least once in historical time, that is to say since human activity has been recorded. These volcanoes are also described as historically active. The second count includes volcanoes from the first category, but also all those that have been active at least once since the end of the last ice age, around 10,000 years ago. Scientists refer to these as Holocene volcanoes.

Mount Kilimanjaro is rated as a Holocene volcano by the Smithsonian Institution. Its neighbour, however, the 5,199-metre-high Mount Kenya, counts as extinct, because there are no traces of an eruption in the last 10,000 years and there is no scientific evidence to suggest that it might erupt again in the future.

The 2,614-metre high volcano Kerimasi in the East African Rift in the north of Tanzania is extinct (right). In contrast, it is immediately evident that the Erta Ale volcano (far right) in the Danakil Desert, northern Ethiopia, is active. A lava lake has been boiling in one of its two craters for around one hundred years. Sometimes the level of the lava lake rises so high that the glowing molten material spills over the edge of the crater.

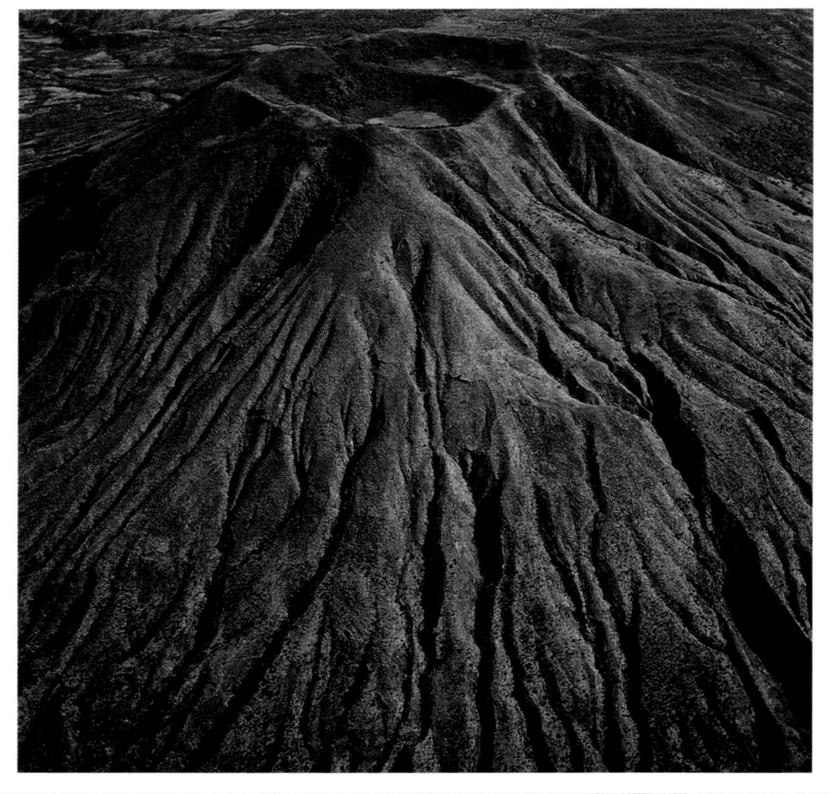

Mount Kilimanjaro – at 5,895 metres the highest mountain in Africa – towers majestically over the scrubland around the Momela Lakes in Arusha National Park, Tanzania. From a distance the gigantic volcano seems to be extinct, but it is only dormant.

Inside a volcano

A volcano basically consists of a magma chamber, a vent through which the magma rises to the earth's surface and a crater through which the magma is expelled. The magma chamber could be described as the volcano's storage cellar. Magma, which forms at great depths, gradually rises up through pores and cracks because it is lighter than solid rock and collects here. The size and shape of the magma chamber, as well as the depth at which it sits, differs from volcano to volcano. These magma reservoirs in the earth's crust are usually found at depths of around three to four kilometres beneath the earth's surface, but can lie as deep as 20 or 30 kilometres. In rare cases the magma chamber sits within the structure of the volcano, not far below the crater. When a volcano erupts, magma is pushed upwards through the vent – which is likely to be more like a network of fissures at its deeper points, only narrowing into a single pipe inside the volcanic cone. Magma then breaks through the crater and is expelled.

This simple basic structure notwithstanding, there are a variety of different forms of volcanoes. The classic version is the stratovolcano – most of the volcanoes on the earth's continents belong to this category, and they are mainly found along subduction zones. The magma from stratovolcanoes is viscous, gassy and often highly explosive. A large proportion of the molten material thrown out during eruptions is expelled in the form of clouds of ash, or pyroclastic flows that pour down the steep sides of the volcano. Only a small amount of magma flows out as lava. These alternating emissions are deposited layer upon layer around the crater. Over time this process creates steep cones that tower majestically over the landscape.

The slopes of stratovolcanoes are, however, not very stable. Particularly violent eruptions, and even small earthquakes or heavy rainfall, can cause the slopes to slide away. This is how the large horseshoe-shaped craters of stratovolcanoes are formed, as well as the recesses that resemble amphitheatres in their flanks. The famous eruption of Mount St Helens, in the Cascade Range, Washington, USA, in 1980, was made worse by such a landslide. Today its active vent lies in a crater that is 2–3.5 kilometres wide and open to one side. From time to time another phenomenon typical of stratovolcanoes forms inside this crater – a lava dome. This structure, chiefly formed in stratovolcanoes with particularly viscous lava, edges up out of the crater by a few metres a day, blocking up the vent like a plug. It is red hot and can grow tens or even hundreds of metres high. If the supply of magma increases or an earthquake

In December 1992, lava flows from the Kilauea volcano on Hawaii's Big Island reached the beaches of Kamoamoa Bay, setting palm trees there on fire (right). Insulated by their own surface crusts, the lava flows on Hawaii can travel for many kilometres. When they reach the sea, many lava flows are still red hot (far right). Layers of lava build up on top of each other. As the coastline gradually pushes forward, the island continues to grow.

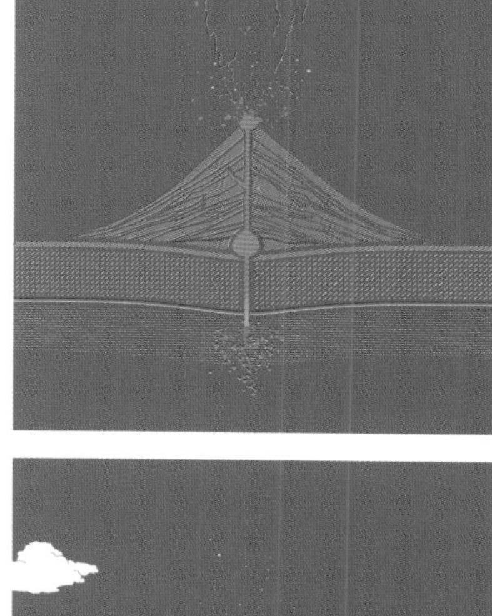

Most continental volcanoes are stratovolcanoes (top left). The steep cone form is typical of stratovolcanoes, while shield volcanoes are extremely flat, like the volcanoes of Hawaii, for example (bottom left). The thin lava of shield volcanoes like Kilauea on Hawaii solidifies into thick, flat pancakes or twisting, rope-like strings (right).

Inside a volcano

If its magma chamber empties during a particularly heavy eruption, a volcano will collapse, leaving behind a cauldron or caldera. It usually takes thousands of years for the chamber to fill up again. Around 7,700 years ago, the caldera of the Karymsky volcano on the Kamchatka Peninsula, Russia, caved in as a result of a massive eruption. Two thousand years later the ground started to rumble again. Over time, further eruptions built up the young cone of the Karymsky volcano, which is 1,486 metres high today (top).

unsettles the lava dome, causing it to break apart, extremely dangerous pyroclastic flows can develop. Since the 1980 eruption of Mount St Helens, two lava domes have formed in close proximity to each other, the second of which stopped growing in October 2008.

In contrast to stratovolcanoes, shield volcanoes have very shallow slopes, so shallow in fact that they are barely recognisable as volcanoes from a distance, even though they can be vast. Mauna Loa on Hawaii, for instance, is the largest shield volcano on earth, as well as the largest active volcano. It rises only 4,165 metres above sea level, but if measured from the ocean floor about 5,000 metres below, it exceeds Mount Everest in the Himalayas, the highest mountain on earth, with a height of over 9,000 metres.

Shield volcanoes expel their magma mainly in the form of glowing lava flows that are low in gases, have a very low viscosity and can flow at speeds of up to 50 kilometres an hour. As the lava flow cools at its edges and on the surface, it forms tunnels. In these tunnels the lava is insulated, can stay liquid for longer and is therefore able to flow for many kilometres before it gradually solidifies at the end of the tunnel. This creates giant, flat mountains. Because their form is reminiscent of a large, round piece of armour lying on the ground, these volcanoes are called shield volcanoes.

Similarly, large amounts of low-viscosity lava flow out of fissure volcanoes. They have no central crater and are fed by elongated clefts that run deep into the earth's interior. This sort of volcano is primarily found on the bottom of the sea along the axis of mid-ocean ridges, but also at spreading zones on land. They often do not leave behind any mountains, just a broad, flat plain of lava. Many recurrent fissure eruptions will form mighty lava plateaus, like the Columbia Plateau in the northwest of the North American continent, which is up to 1,800 metres thick.

Sometimes, however, chains of craters made up of either ash or chunks of lava do develop along fissures – these are called cinder cones and tuff cones. They can also form on fissures in the flanks of shield volcanoes and stratovolcanoes. When shield volcanoes and stratovolcanoes erupt they can lose so much magma that the roof of the magma chamber under the volcano caves in. This can cause the volcano to completely collapse, particularly in the case of stratovolcanoes. Usually a circular hollow – a caldera – is left behind after this sort of disastrous eruption. The caldera often fills up with ground- and rainwater, and a new volcanic cone can rise out of it over time.

A young forest is starting to spread out in the crater of this cinder cone (right). There are 400 craters of this type along the ridge and in the caldera of the Newberry shield volcano in Oregon, USA. The cone of Gorely, a stratovolcano on Russia's Kamchatka Peninsula, is made up of many layers of ash and lava (far right). The crater walls cut deeply into the different layers, and an acid lake bubbles on the crater floor.

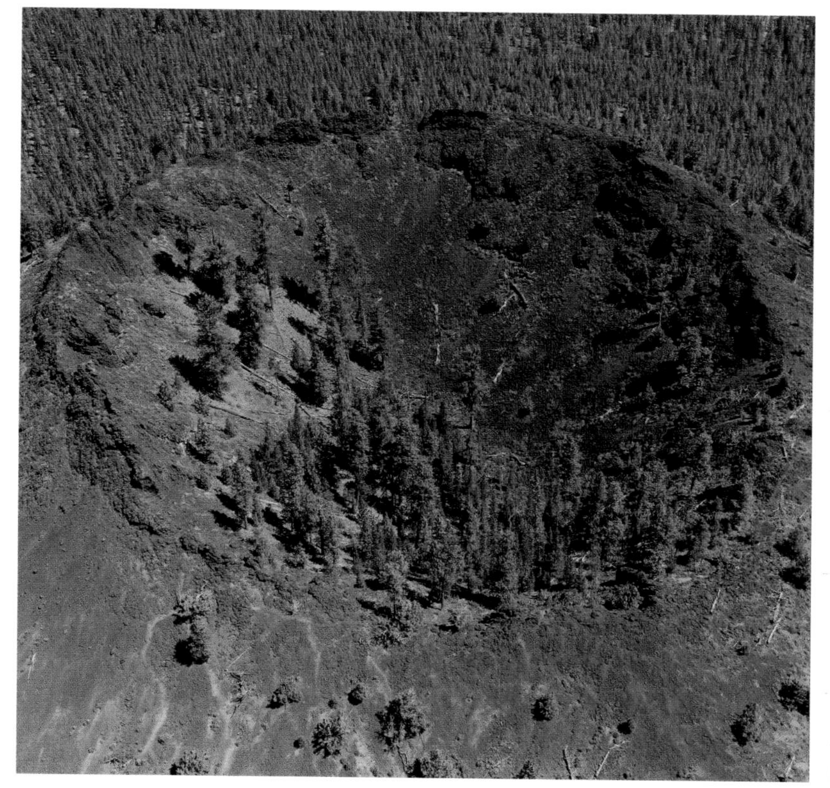

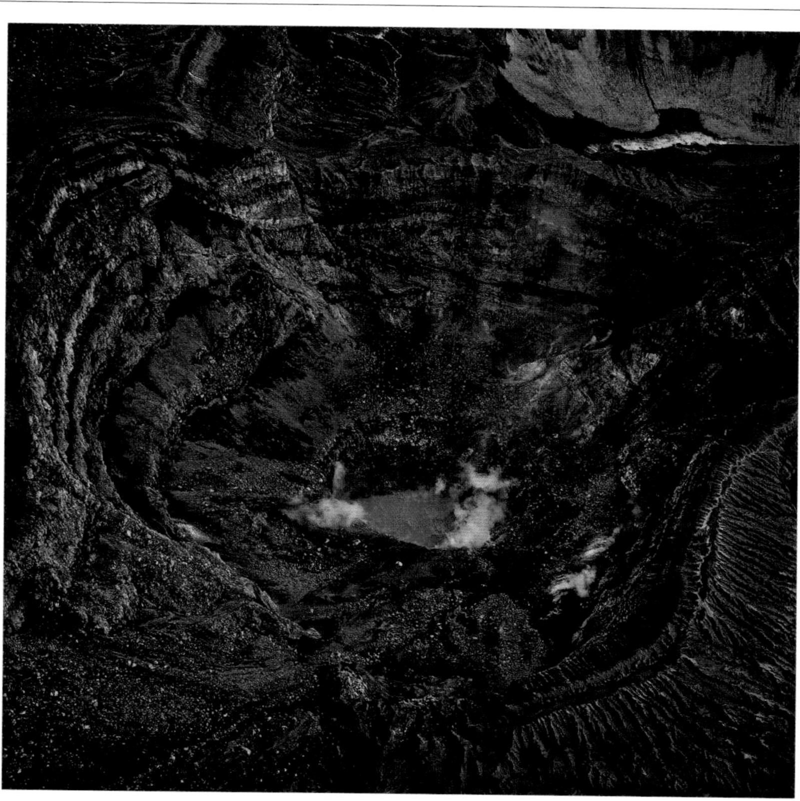

The life of an active volcano

Although active volcanoes can be divided into specific categories (see page 188), every volcano behaves in its own individual way, which makes predicting eruptions difficult. Particularly dangerous volcanoes are therefore not only observed in detail over many years, but are also scrutinized for clues that allow their life stories to be reconstructed. The better scientists know a volcano and the more information they have about earlier eruptions, seasonal progressions, the strength and volume of ejections, seismic activity in the bedrock or the chemical composition of volcanic gases, the earlier they can warn of a potential eruption.

The most dangerous of Europe's volcanoes is Vesuvius, which rises 1,281 metres above the densely populated Bay of Naples. It lies near the subduction zone between Africa and Europe and is highly explosive, like all subduction volcanoes. Its most destructive historical eruptions compare with the eruption of Mount St Helens in North America's Cascade Range in 1980, and the 1991 eruption of Mount Pinatubo in the Philippines.

Scientists believe that volcanic activity in the area over which Vesuvius now looms began some 300,000 years ago, under the sea. Around 25,000 years ago, Monte Somma, the precursor of Vesuvius, rose out of the sea as a volcanic island and gradually fused with the mainland as it grew higher. Abundant layers of volcanic rock found more than 10 kilometres away from the site of the current volcano suggest that Monte Somma must have had five devastating eruptions, with the first four ocurring 18,000, 17,000, 8,000 and 3,360 years ago and the most recent in AD 79, when the cities of Pompeii, Herculaneum and Stabiae were completely buried by falling ash and pyroclastic flows.

The eruption of Vesuvius in AD 79 is the first volcanic eruption in human history to have been documented in detail. In two letters to the Roman historian Tacitus, Pliny the Younger described how his uncle, the natural historian Pliny the Elder, who was staying in Stabiae at the time, perished during the eruption. It is believed that Pliny, undaunted, went to visit a bath house as residents were already beginning to flee the city. Earthquakes were shaking the ground and the first clouds of ash were erupting out of the crater. When Pliny finally decided to leave Stabiae, the streets were already dark with falling ash and people were panicking, tying pillows to their heads to protect themselves from falling rocks. It is likely that Pliny collapsed in the street and suffocated in a cloud of hot gases. His body was recovered after the eruption.

During this eruption, Monte Somma collapsed and a caldera was created, in which Vesuvius slowly grew. Like a shirt collar, the edge of the old Somma caldera half encircles the young cone, which has erupted more than 20 times over the last 300 years alone. The last major eruption took place in 1944, towards the end of World War II. As the aeroplanes of the Allied forces were targeting the Axis troops, Vesuvius fired giant clouds of ash into the sky, and 12,000 people were evacuated. Although it was not a particularly violent eruption, 26 people perished and 88 fighter aircraft on the ground were destroyed.

There has been no further activity since then – no ash thrown out, no lava flows. Only slight tremors and light gas hazes that stream out of the 100-metre high crater walls attest to the fact that Vesuvius is not extinct, only sleeping. The reconstructions of its history show that the longer it sleeps, the more violent its next eruption is likely to be. Today, over a million people live in densely populated, built-up areas directly within the danger zone. Although scientists keep track of the volcano's every movement via a network of monitoring stations, it is unclear whether the area could be evacuated quickly enough in an emergency.

There are volcanoes in the middle of the European continent that have spat out lava and ash in the last 10,000 years and therefore still count as active: in the Massif Central in south-central France, for example, and also in the Eifel region of

One of the most dangerous volcanoes on earth, Mount Vesuvius rises 1,281 metres over the Bay of Naples, Italy. Hundreds of thousands of people live in the surrounding area.

western Germany. There there are hundreds of eruption points in what is now a landscape of gentle, green hills, but very few craters have been preserved here.

The Eifel volcanoes are fed by a mantle plume that reaches up to around 40 kilometres below the earth's surface. Magma has risen up from there many times over the last 50 million years and broken through the earth's surface. Volcanologists have confirmed over 100 eruptions during the last 700,000 years, with quiet phases between eruptions lasting between 10,000 and 20,000 years.

The most destructive volcanic eruption in the Eifel took place around the end of the last ice age, approximately 12,000 years ago, and was similar in strength to the eruption of Vesuvius in AD 79. Clouds of ash rose many kilometres into the air and pyroclastic flows rushed down the flanks of the volcano all the way to the valley of the Rhine, one of the largest rivers in Europe. They spilled into the riverbed, damming up the water and creating a 20-kilometre long lake in the Rhine valley. After only a few days, however, the Rhine broke through the barrier and continued along its original path. About 40 kilometres southwest of this point, the most recent volcanic eruption in the Eifel took place, around 9,000 years ago. Molten lava forced up through the bedrock met with groundwater and caused giant steam explosions, which in turn blasted out the Ulmenmaar. It is a crater lake today.

According to geophysical measurements, the temperature in the bedrock of the Eifel is higher than is normal in the earth's crust and the landscape is rising up by around 1 millimetre a year, which suggests that magma is being pushed up from below. Carbon dioxide also streams out of the ground in many places; this, we know from chemical analyses, comes from deep in the earth's mantel. The area is kept under observation because of these factors but, according to scientists, the next volcanic eruption in the Eifel is unlikely to happen for another few thousand years.

The volcano on the southern Italian island of Stromboli is famous for its constant activity (right). Several times a day, frequently several times an hour, it spits out glowing lumps of lava in spectacular fountains up to 100 metres high. It has done so throughout recorded history, which is why sailors in ancient times called it the 'Lighthouse of the Mediterranean'. The Pulver Maar is one of 240 eruption centres in the Eifel region in western Germany, where there were active volcanoes until 9,000 years ago (far right). It has a diameter of 651 metres and is 72 metres deep.

Pyroclastics

Following the principle that 'there's no smoke without fire', people used to believe that volcanoes acted as chimneys for fires burning in the earth's interior. They therefore identified the fine sand and dust ejected from craters during eruptions as ash, and believed that the glowing pieces of lava that flew through the air and landed on the flanks of the volcano, solidifying into bizarre lumps, were scoria. At the end of the nineteenth century, naturalists put the dark granules under the microscope for the first time and established that they were not, as they had imagined, looking at burnt remains from subterranean coal deposits, but instead at tiny, vitreous fragments of stone.

Even today, volcanologists use the term volcanic ash for any material that comes out of a volcano's vent in an eruption cloud and is no larger than two millimetres in diameter. Larger particles are called lapilli, an expression that originated with Italian quarrymen and basically means 'little stone'. This is a little misleading, as pieces the size of walnuts also count as lapilli. Larger lumps of lava that fly from the volcanic vent glowing red-hot and solidify, before finally plummeting to the ground some distance from the crater, are often referred to as bombs.

During heavy eruptions large rocks are also torn away from the walls of the vent and swept along, as famously occurred during the eruption of the Arenal volcano in Costa Rica in 1968. When the volcano awoke after lying dormant for hundreds of years, it threw rocks weighing tonnes out of its crater, some of which crashed to earth five kilometres away. According to scientific calculations they must have been catapulted out of the crater at a speed of 2,000 kilometres an hour in order to cover this distance. On impact, the large rocks from the Arenal volcano blasted open 60-metre wide pits that ripped several metres into the ground. This terrible hail storm turned an area of more than 10 square kilometres around the volcano into a battlefield.

Made up of the Greek words *pyros* (fire) and *klasis* (to break), the term 'pyroclastic' is used by scientists to categorize volcanic ash, lapilli, cinders, bombs and blocks of stone – in fact anything that is catapulted into the air during an eruption. Pumice, the rock filled with countless tiny holes that is light enough to float on water, also belongs to this category. Pumice is created when gas-rich magma foams as it is pushed out of a volcanic vent. These foam fragments are thrown high out of the crater and solidify before they fall to the ground.

Volcanic eruptions do not, however, always rise vertically into the air. If they are too heavily loaded, they simply spill over the edge of the crater and race down the flanks of the volcano as

Giant lumps of lava, so-called breadcrust bombs, and huge quantities of rubble can be found in the area surrounding Arenal, a 1,657-metre high volcano in Costa Rica (right). They were thrown out of the crater during a major eruption in 1968 and landed many kilometres away. Layers of ash and pumice from various eruptions show that there were active volcanoes in the Eifel region of Germany during a recent stage of the earth's history (far right).

The steaming lava dome of the 3,772-metre high Santa María volcano in Guatemala is called Santiaguito. A massive eruption in 1902, one of the worst of the twentieth century, tore away part of the volcano's flank. Several lava domes have grown here since 1922. The youngest, Santiaguito emits blocky, viscous lava as well as clouds of gas and sometimes ash.

The ruins on the flank of the Soufrière Hills volcano on the Caribbean island of Montserrat are evidence of the destructive power of pyroclastic flows. Made up of hot rubble, ash and corrosive gases, these glowing avalanches have temperatures of several hundred degrees Celsius. When the pressure in the volcano's lava dome becomes too powerful, they roll down the mountain, burying what still remains of these villages.

a grey avalanche of dust, reaching speeds of up to 500 kilometres an hour. The gases that expand and escape from the superheated mixture play important roles here, as they make the cloud highly mobile. Scientists call this phenomenon pyroclastic flow or – from the French – *nueé ardente*.

When subduction volcanoes erupt they often produce pyroclastic flows as well as ash clouds. This is what makes them so dangerous. These glowing clouds scorch everything in their path – anyone caught up in them has no chance of survival. On the small, mountainous island of Montserrat in the Caribbean, British scientists were able to study pyroclastic flows from a safe distance. In July 1995 the old Soufrière Hills volcano erupted for the first time, after a dormant phase lasting hundreds of years. Pyroclastic flows devastated Plymouth, the capital city, as well as the southern part of Montserrat. Following the eruptions, 8,000 of the 12,000 inhabitants of this tropical paradise left the island, and 4,000 moved to its safer northern part.

Since then, the volcano and its pyroclastic flows have been constantly monitored. Scientific studies have shown that pyroclastic flows are always prefigured by a characteristic earthquake signal, so as soon as this shows up on the instruments, the island authorities are notified and the population warned.

Thick ash clouds shot out of a new crater on the flanks of Mount Etna in Sicily, Italy, for many months in 2002. The eruption turned Etna's summit, an area previously marked by lava flows, into a desert of ash.

There have been further discoveries: the pyroclastic flows on Montserrat can reach temperatures of 600 degrees Celsius, speeds of 200 to 300 kilometres an hour and leave behind deposits at the foot of the volcano, often metres high, which retain volcanic heat for a long period of time. Even seven months after the initial eruption in July 1995, scientists measured temperatures of 350 degrees Celsius two metres below the surface of these fresh deposits.

Pyroclastic flows occur on the Soufrière Hills volcano whenever the lava dome on top of the volcano collapses. Lava domes are another typical phenomenon of the highly explosive volcanoes at subduction zones. Hot mounds of debris form directly over the opening of volcanic vents and can rise up hundreds of metres into the air. They are created when extremely viscous magma is squeezed bit by bit out of the pipe in the crater, like toothpaste out of a tube.

Fresh lava rock often breaks into pieces the moment it leaves the crater, and steaming rubble then rumbles down the sides of the dome. However, sometimes slender, hot lava spines grow many metres into the air before collapsing. At times, 13 cubic metres of viscous lava per second were pushed out of the vent on Montserrat in 1995, and the spines reached heights of up to 70 metres before they collapsed. The release of pressure after such a collapse results in heavy explosions that can destroy a whole dome. A hot mixture of rock pieces, shreds of lava, volcanic ash and corrosive gases – a pyroclastic flow – surges down the slope of the volcano, through a river valley and over the ruins of the former villages on its way to the sea.

It is rare to see streaming, red-hot lava coming from the highly explosive volcanoes of subduction zones. If lava flows at all, it is viscous and solidifies into sharp-edged boulders. Liquid lava that flows out at high speed, running down the sides of volcanoes like glowing rivers, is only found at shield and fissure volcanoes.

Risks and benefits

Around 500 million people worldwide live in the vicinity of active volcanoes – around a tenth of the world's population. Those who live near active volcanoes expose themselves to risk: the earthquakes that accompany eruptions can set off landslides, cause tidal waves and damage or completely destroy houses. Pyroclastic flows and mudslides burn and destroy everything in their paths. Falling ash can bury houses and cause roofs to cave in. Plants die away under thick layers of ash, leading to failed harvests and the possibility of famine, while volcanic dust and corrosive gases cause breathing difficulties. According to statistics, five per cent of all volcanic eruptions lead to fatalities. In the last 300 years around 260,000 people have been killed worldwide as a result of eruptions.

Many people consciously put up with these risks, however, because they can not only profit from volcanoes, their livelihoods can even depend on them. Many people benefit from the fertility of the soil in volcanic areas. As a result of weathering over time, the discharge from volcanic eruptions in humid, warm regions turns into mineral- and nutrient-rich soil that is generally loosely packed and therefore aerated and easy to work. Furthermore, the tiny pores in volcanic ash hold moisture. The farmers of the arid, rain-starved island of Lanzarote in the Canaries turn this to their advantage. Their vines and fig trees flourish in the black volcanic soil, because the tiny pores in stones spat out by the island's volcanoes suck in moisture during the night, which they slowly release into the roots of the plants during the following day.

The bedrock in areas with volcanoes active and extinct lends itself particularly well to the extraction of geothermal energy. Countries that use geothermal energy to meet a large proportion of their electricity and heating needs include Japan, New Zealand, Mexico, Indonesia, the Philippines and, within Europe, Italy and, notably, Iceland. Eighty per cent of homes on this Atlantic island are heated with geothermal energy, as are greenhouses in which various vegetables and even bananas thrive. Twenty per cent of Iceland's electricity comes from power stations whose turbines are driven by steam from the earth.

Volcanic rock is also important for the construction industry. Basaltic lava, ignimbrite and tuff are popular building materials in volcanic regions, and buildings such as Germany's famous Cologne Cathedral were built of trachyte gathered from nearby extinct volcanoes. Basalt is used to make cobblestones, while pumice stone is used in the manufacture of lightweight concrete, as well as serving its popular purpose as a smoothing and cleaning material.

Every one of the little hollows in the volcanic soil of Lanzarote, one of the Canary Islands, contains a fig tree or a vine (right). The soil stores up moisture over night and slowly releases it during the day, while circular walls made out of lava rocks protect the plants from the desert wind. Until the early twentieth century, the people living in the villages at the foot of the 926-metre high volcano on the Aeolian Island of Stromboli, Italy, grew olives, capers, vines and vegetables on the steep slopes (far right). These days, their main source of income is tourism.

Ancient, long-since extinct volcanoes also hold ores. Iron, zinc, copper, lead, tin, tungsten and molybdenum, even gold and silver, are found in the pores and seams of rocks near cooled magma chambers – sometimes just in small traces, sometimes in mineable amounts. In order to explain why ore is found in these places, geologists like to make a comparison between magma chambers and giant distilling flasks. While the volcanoes were active, corrosive liquids – hot fluids, rich with metallic compounds – were discharged from the molten material, and penetrated the solid layers forming the 'roof' of the magma chamber. There they circulated in gaps and narrow fissures in the stone and reacted with the solid rock, causing the minerals in this rock to dissolve. The fluids also often mixed with sea- or rainwater that had seeped down from the surface. As soon as these hot mixtures reached the cooler areas of rock higher up, where there was less pressure, they deposited their load of solutes, among them metallic compounds. Geologists therefore often scrutinize the base rock of old volcanoes when they are on the hunt for new ore deposits.

Ocean waves gently wash over the snow-covered, black lava sands on the south coast of Iceland. The inhabitants of this volcanic island in the North Atlantic make use of the underground heat to generate electricity and provide central heating.

Plate tectonics: spreading

From space, the East African Rift and its neighbouring mountain chains look like a gigantic, wrinkled scar, cutting into the earth's crust to a depth of around 1,000 metres. To the north, at the Afar Depression, it spans a massive 570 kilometres, but narrows as it travels south to between 50 and 200 kilometres before finally splitting into two branches encircling Lake Victoria, the largest lake in Africa.

Earthquakes as well as volcanoes have an impact on the landscape surrounding this rift. According to contemporary thinking, the reason for this underground unrest is a row of hot magma masses, or mantle plumes, that flow upwards from the earth's interior. These push against the continental crust, causing it to bulge. As a result, the earth is lifted up and stretched to such an extent that it tears. Fault blocks then collapse along the axis of the bulge, and a trench develops. Red-hot molten rock pushes up through cracks and bursts out through fissures and volcanic vents.

This powerful process, which scientists call 'spreading', began in the East African Rift around 25 million years ago. The most intense activity occurred between 2.5 and six million years ago, though the process continues today at a slower rate. According to satellite measurements, the Afar Depression widens by up to 12 millimetres every year, with the narrower areas of the Rift Valley in southern East Africa growing by an annual average of six millimetres. This seems quite modest until one realizes that, in only 1,000 years, the Rift Valley will have expanded by 12 metres in the north and six metres in the south.

Traces of the massive underground forces at work in this region can best be seen in the barren deserts of the Afar Depression. The earth's crust is completely torn here, where there is a deeply sunken area three times the size of Switzerland. The deepest area – Lake Assal, in the heart of Djibouti – lies around 155 metres below sea level.

The desert floor, covered by cooled lava, is scarred with countless fissures, often studded with small volcanic cones. Faults many kilometres long cut the earth's crust into blocks which lean against one another like steps, creating an imposing fault scarp landscape.

Through the Afar Depression, the East African Rift is connected to an even larger spreading zone – the network of mid-ocean ridges that runs beneath oceans and seas around the world. Here, two parts of this network – the ridge which runs along the bed of the Red Sea and the Carlsberg Ridge in the Gulf of Aden – merge with the East African Rift in what geologists

Erta Ale is a shield volcano with an elongated caldera on its ridge, located in the Afar Depression of northern Ethiopia. The very fluid lava that streams out of its pit craters during eruptions is typical of spreading zones. The illustration shows the tectonic situation at the Horn of Africa: the Erta Ale volcano lies in the north of the East African Rift (dotted lines), thus in the spreading zone where Africa threatens to break in two – into the Somalian and the Nubian Plates. The white arrows indicate the direction of movement of the individual plates, while triangles indicate the position of volcanoes in the region.

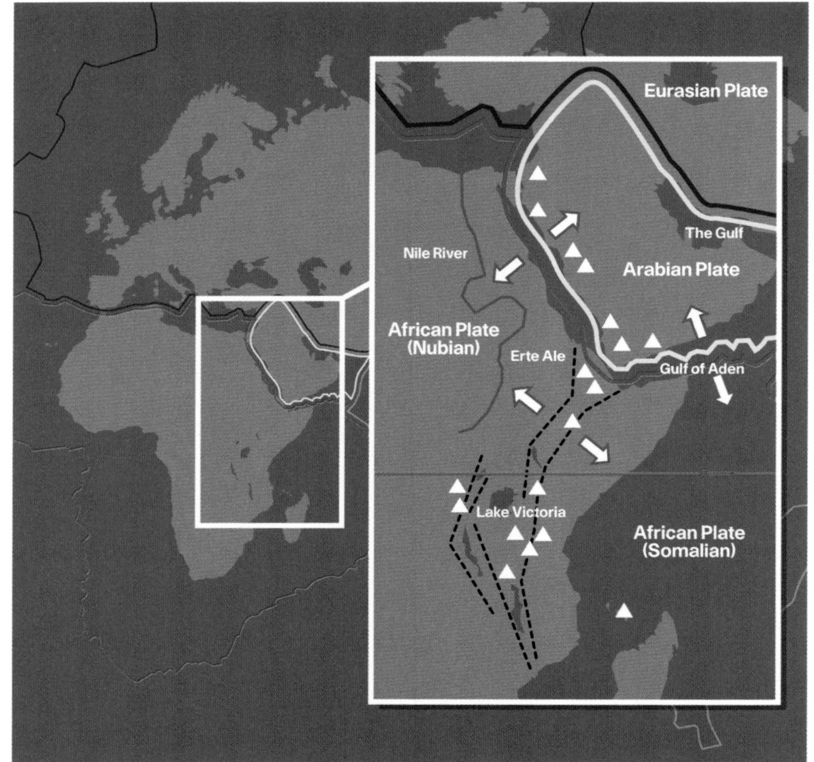

call a ‘triple point’. The result is a star-shaped formation in which the individual spreading zones pull apart from one another at 120-degree angles.

The spreading occurring along the bed of the Red Sea and in the Gulf of Aden began around 10 million years earlier than that in the present-day East African Rift region, so the trench running through the two regions is much more developed. A section of the African continent has already completely broken off – the Arabian Plate. Magma is constantly flowing up and pushing through cracks in the sea-floor, forcing Africa and the Arabian Peninsula further apart, at a speed of up to 19 millimetres a year.

Similar plate-tectonic relationships to those on the floor of the Red Sea and in the Gulf of Aden, though less advanced, exist in the Afar Depression, making this hot, hostile desert an ideal environment for geologists. They can keep their feet dry by studying phenomena on land that would otherwise require a great deal of costly diving equipment to investigate deep beneath the sea – namely, how emerging magma is able to tear and push apart the hard, rocky crust of our planet; how plates start to drift apart; and how a mid-ocean ridge is gradually formed. The only other place in the world where one can observe a similar spreading process is on the Atlantic island of Iceland.

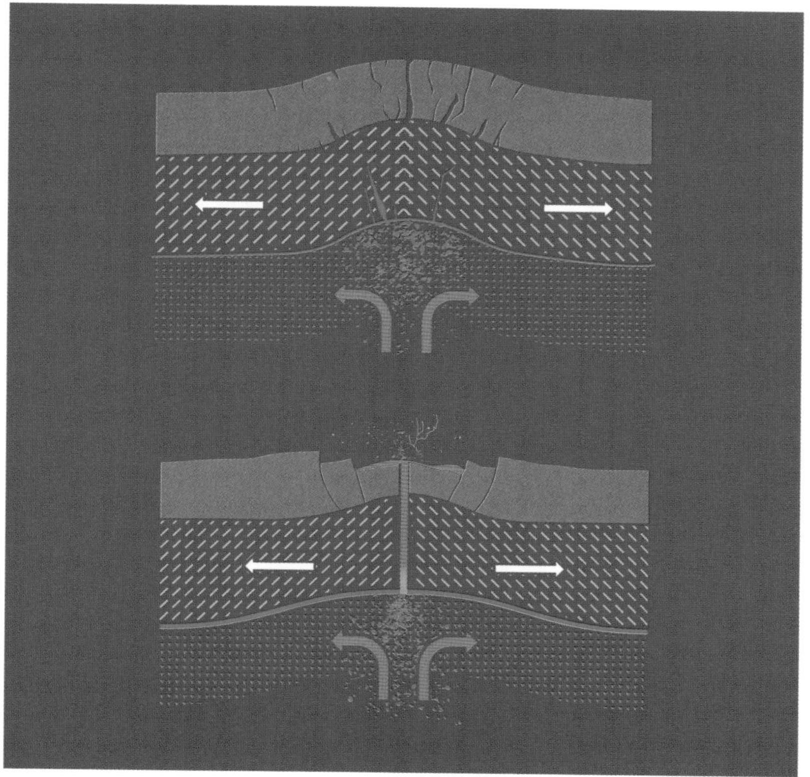

Lake Magadi, with its surface covered in floating salt crusts, also lies in the middle of the East African Rift (above). Its western shore directly borders one of the fault planes where the earth's crust has broken open due to the spreading process, creating stepped formations. The scarp is over 100 metres high. The schematic illustrations (above right) show the spreading and ripping of the African continent in the East African Rift region in two cross-sections.

Almost parallel, kilometre-long faults cross the lava desert around Sac Allol in northern Djibouti, splitting it into massive, staggered blocks. Scientists call this a scarp fault topography. This landscape is typical of continents where the earth's crust spreads. Following rare rainfalls, muddy water collects in the elongated cavities formed when blocks of the earth's crust subside.

Lava that has been pushed up out of the fissures in the Afar Depression over the last 2 million years is similar in its chemical composition to the lava emerging from submarine mountain ranges. Indeed, up until 65,000 years ago, saltwater sometimes flowed into the region, turning the Depression into a side basin of the Red Sea. The water quickly evaporated each time, depositing a large quantity of salt in the basin, which the people of Afar now mine and transport away by camel caravan. There were also volcanoes in this now dried-out side basin of the Red Sea. Today the stumps of these once-undersea volcanoes rise up over the dry, hot desert floor. The remains of coral limestone can be found on their flanks.

The African continent has yet to break apart along the East African Rift and scientists are not in agreement as to when or even if this will ever happen. Yet they have already given a name to the section of the continent that might break away over the course of the next few million years – the Somalian Plate. If the hot magma flowing upwards from the earth's interior should dry up, the spreading movement in this region would cease and the African continent would keep its current size.

On the western side of Africa, in the area around present-day Namibia, there is evidence to suggest that processes similar to those occurring today in the East African Rift were at work 130 million years ago. Magma pushing up through the earth's crust caused the South American Plate to separate from the African Plate and the Atlantic ocean began to form. In the Damara region of Namibia, elongated ridges and circular domes of hard, dark rock rise up over the desert floor, bearing witness to the volcanic activity of this period. These are the remains of old lava flows from long-extinct and eroded craters as well as solidified magma that never reached the earth's surface but got stuck and cooled underground. Weathering and erosion have since exposed these formations, which include Brandberg Mountain, the Spitzkoppe and Mount Erongo.

The dark rocky hills of the Messum Crater in the desert of Namibia are the remains of a volcano, now long extinct, that erupted here around 130 million years ago (right). At that time the earth's crust on the ancient supercontinent of Gondwana began to spread and rip apart, at which point the South American and African Plates were created, with the Atlantic ocean between them. Stemming from the same period, the granite rocks of the Spitzkoppe rise up around 100 kilometres to the east (far right). They were created from rising magma which cooled and solidified in the earth's crust before reaching the surface, only to be freed much later through erosion and further shaped by weathering.

Plate tectonics: subduction

Wherever tectonic plates collide, vast forces set the earth's crust in motion. Mountain ranges are pushed upwards and chains of volcanoes are created along the subduction zones, where one plate dives beneath the other. If two plates with oceanic rather than continental crusts come into contact and one is pushed beneath the other, volcanic islands grow out of the sea as a result, often arranged in an arc, like a string of pearls.

The eruptions of volcanoes along subduction zones are usually highly explosive. The primary reason for this is the large amount of water held in the marine sediment and the pores of rocks in the lithosphere, that is dragged into the depths of the earth with the subducting plate. At a depth of around 50 to 150 kilometres, the increasing pressure and heat ensure that the water is forced out, and the plate gradually melts. Instead of melting at the normal temperature of around 1,000 degrees Celsius, the solid rock can start to liquefy at lower temperatures due to the extremely hot water in the subduction zone. The high pressure intensifies this process. Magma forms.

Because the molten sections are lighter than the solid rock in the surrounding area, the magma rises up through tiny cracks and holes, like drops of oil in a glass of water. On its way up, more solid rock is melted. The drops of magma, which move through the solid rock at a rate of just 30 centimetres a year in some places and 50 metres in others, gradually gather and collect a few kilometres beneath the earth's surface in large magma chambers. The magma chamber is the engine of a volcano, and normally lies only three to four kilometres beneath it. Magma is pushed up into the vent from here and rises up to the crater opening. When the pressure in the magma chamber increases, shortly before an eruption, the volcanic cone above it often swells up, and the earth begins to tremble. Scientists therefore regularly measure the gradient of a volcano's slopes and keep track of earthquakes in the area – both are key indications of an impending eruption.

The molten rock of subduction zone volcanoes is usually very viscous. The gas content is also very high; alongside sulphur compounds, this mainly consists of carbon dioxide and steam. At great depths these gases are dissolved in the magma due to the high pressure. As the magma climbs, the pressure decreases and the gas can escape, creating bubbles that make the magma accelerate on its way up. On reaching the crater opening, the steam and gas can suddenly discharge. This results in huge explosions: the magma is blown to pieces and, during particularly heavy eruptions, the finest particles, volcanic ash and dust, can be catapulted up to 50 kilometres into the atmosphere.

A 'volcanic storm' over the cone of Krakatau in Indonesia: the electricity that builds up between particles thrown out of the crater during an explosive eruption is discharged in bolts of lightning. Krakatau lies in the subduction zone between the Indoaustralian and Eurasian Plates.

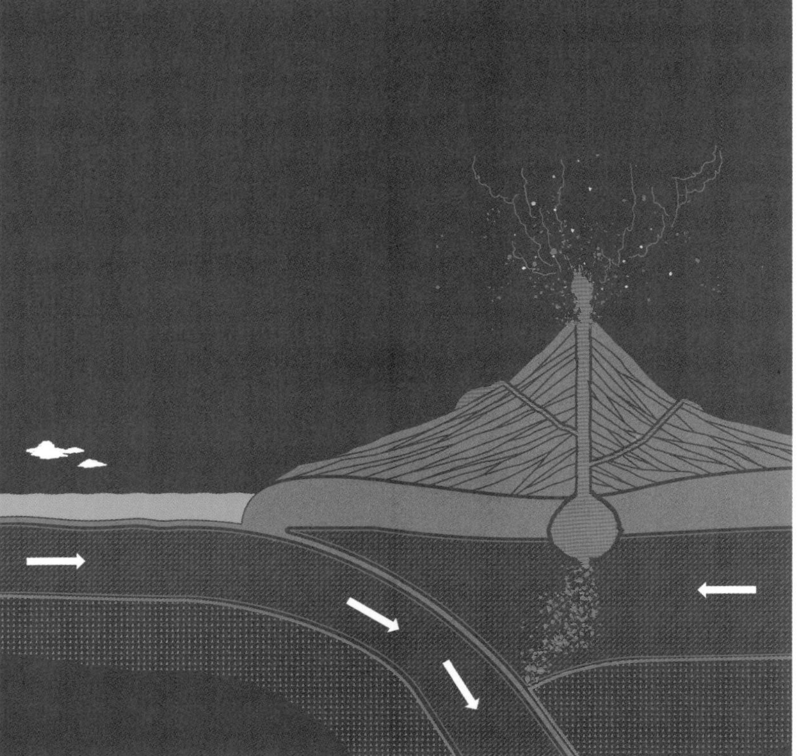

Stratovolcanoes with their steep cones are typical of subduction zones, where one plate dives under the other and melts (illustration opposite). On the coast of Kamchatka, where the Pacific Plate dives under the Siberian spur of the North American Plate, the 4,750-metre high Kliuchevskoi rises up, the highest active volcano in Asia (right).

Supervolcanoes

In March 2005, a study by scientists from the British Geological Society highlighting the potential danger of supervolcanoes attracted worldwide attention. According to this report, the eruption of such a giant would be comparable to the impact of a kilometre-wide meteorite. It would have serious consequences for the entire planet and could threaten the whole of humanity. The primary danger would come from volcanic gases, which would be blown up to 50 kilometres into the air and spread around the planet, thereby significantly changing the global climate for a number of years. The resulting 'volcanic winter' would cause widespread famine. Concluding their report by saying that a super eruption of this type will happen sooner or later, the scientists called for an increase in research on supervolcanoes, including observation, the education of the world population and the establishment of a multi-disciplinary task force that would coordinate activities internationally in the case of such a catastrophe.

Supervolcanoes are defined by the amount of magma they expel over a short period of time during an eruption. This must amount to at least 1,000 cubic kilometres, 10 to 100 times more than the volume emitted during a 'normal' volcanic eruption. During the devastating eruption in Indonesia in 1883, Krakatau emitted around 10 cubic kilometres of magma. In the two most devastating eruptions of the last 50 years, in 1991 and 1980 respectively, Mount Pinatubo in the Philippines ejected 4.8 cubic kilometres of magma and Mount St Helens in the North American Cascade Mountains 'only' 0.4 cubic kilometres.

In contrast, the world's best-known supervolcano, the Yellowstone volcano, which currently lies dormant beneath the state of Wyoming, drove 1,000 cubic kilometres of volcanic material into the atmosphere 640,000 years ago. During an eruption 2.1 million years ago, it was 2,450 cubic kilometres of material. The sheer quantity of material released can be explained by the fact that the magma chambers of supervolcanoes are very large. The chamber underneath Yellowstone National Park is estimated to be around 60 kilometres long, 40 kilometres wide and 10 kilometres deep. It lies around five kilometres below the earth's surface and is fed by a mantle plume.

Supervolcanoes are also distinguished by the fact that they do not create volcanic cones. Instead their devastating eruptions create a caldera, as the ejection of such huge amounts of molten material and gas in such a short amount of time always causes the roof of the magma chamber to cave in. The Yellowstone caldera, dating back to the last super eruption 640,000 years ago, is 80 kilometres long and 55 kilometres wide. Since then, there have been another 30 smaller eruptions here, during which a large amount of lava flowed out and gradually filled up the hollow.

Even though no supervolcanic eruption is expected in Yellowstone National Park for the next few thousand years, the area is under continual observation. According to satellite measurements, the ground there sometimes rises and sinks by a few millimetres, and the hot springs and geysers that bubble up here serve as an indication of the heat below.

Statistically speaking, supervolcanic eruptions occur on our planet every 50,000 years. So far mankind has survived two of these gigantic eruptions. Around 22,000 years ago Taupo in New Zealand catapulted almost 1,200 cubic kilometres of volcanic material into the atmosphere. Toba, a volcano on the Indonesian island of Sumatra, shot roughly twice as much material out of its vent. Both these eruptions left behind giant craters, now filled by lakes.

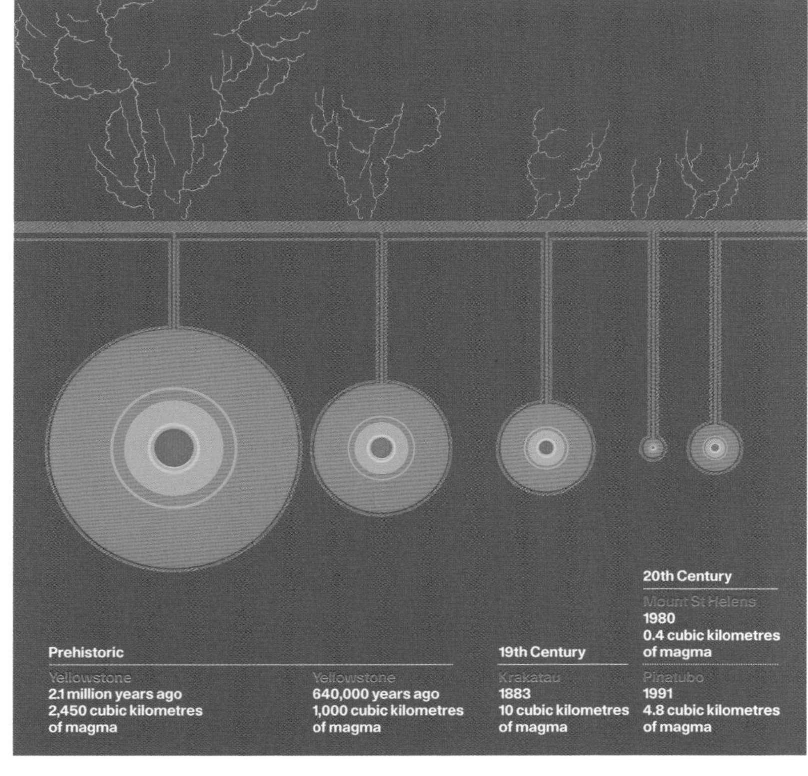

The orange circles represent the amount of matter ejected during various volcanic eruptions, comparing the largest eruptions of the Yellowstone supervolcano with famous volcanic eruptions of the nineteenth and twentieth centuries. This illustration is based on data from the US Geological Survey.

The eruption that blew the summit off Mount St Helens in May 1980 was a small gust compared to the eruption of a supervolcano. However, recent measurements indicate that the magma chamber lying 15 kilometres below this volcanic giant could be connected to those of its neighbours, Mount Rainier and Mount Adams, 70 and 50 kilometres away respectively. If this is indeed the case, this magma reservoir would be larger than that under the well-known Yellowstone supervolcano, and would make Mount St Helens the 31st supervolcano on earth.

Volcanoes and water

Whenever volcanic heat meets water, the earth begins to boil. As if the volcanoes were breathing and sweating, hot vapours stream out of the holes and fissures of its rocks, while mud pools and hot springs bubble. Often these hydrothermal phenomena are a sign that a volcano is simply resting, but not extinct forever.

Fumaroles are plumes of gas that can have a temperature of hundreds of degrees Celsius and often shoot out of the ground at high pressure with a loud hiss. They usually consist of up to 90 per cent steam. The rest is carbon monoxide and carbon dioxide as well as various sulphur and halogen compounds, among them aggressive substances such as hydrochloric and hydrofluoric acid. On their way out of the earth's interior, they dissolve minerals out of the rock so, in places where the clouds of gas condense, they often leave behind colourful deposits and crusts. Sulphur builds carpets of fine bright yellow and orange crystals. Chlorides are white, while iron compounds are yellowish and red to rust-coloured. Copper compounds produce green patches. Over time, volcanic vapours can completely eat away the bedrock: hard rock becomes soft, sludgy clay and pools of water are turned into bubbling, muddy brews.

It's best not to get too close to these phenomena. The crusts are generally unstable, you could break through and sustain burns – and the air near gas vents is also corrosive. Hydrochloric acid and sulphur compounds aggravate the mucous membranes in the mouth and nose. The eyes begin to water. Hydrogen sulphide produces the penetrating smell of rotten eggs. Most poisonous is hydrofluoric acid, which can even dissolve glass. Gas masks are essential for scientists who want to study volcanic vapours on location.

These corrosive gases are also the reason why lakes often form in the craters of dormant volcanoes, as they turn the broken, brittle rock into a watertight mass as part of their corrosion process. Rainwater that falls into the crater can no longer drain off or seep away; instead it accumulates and creates a crater lake. Finally the water turns into acid, due to the sulphuric gases that continue to stream out of the crater floor. Depending on their acidity, mineral content and the quantity of saturated clay, such acid lakes can develop a deep grey, turquoise, emerald-green or rust-brown colour.

The less hot, particularly sulphur-rich vapours, with temperatures of between 100 and 200 degrees Celsius, are called solfataras; those under 100 degrees Celsius are called mofettes. Mofettes can be more dangerous to people than corrosive fumaroles and solfataras because the gas that flows out of them

The fountain of the Prince of Wales Feathers geyser shoots diagonally into the sky first, serving as a prelude to the eruption of the neighbouring Pohutu geyser (right). This performance, from the twin geysers in the Whakarewarewa geothermal region on New Zealand's North Island, is repeated around 20 times a day. Warbrick Terrace also lies on the North Island, in the Waimangu Valley (far right). White, green and ochre algae able to withstand temperatures of around 50 degrees Celsius live in the hot water.

Hot volcanic gases can corrode rock and turn ponds into bubbling mud pools. An upsurge of hot gas causes mud eruptions in the Námafjall geothermal area in the north of Iceland.

cannot be smelled, tasted or seen. It consists of almost 100 per cent carbon dioxide, which is heavier than air and therefore displaces oxygen at ground level. When people or animals get caught in a cloud of carbon dioxide, they suffocate.

A tragic case occurred in 1979 on the Dieng Plateau in the centre of the Indonesian island of Java. The volcanic soil of the highlands, 2,000 metres above sea level, is very fertile, and the heavily populated area is also well known for its complex of Hindu temples, which date from 400 BC and are the oldest on Java. The volcanic cones on the Dieng Plateau have not expelled lava or ash for thousands of years. There are, however, hot springs there, fumaroles and solfataras hiss out of the ground, and grey mud bubbles in a few old crater holes. Every few years, phreatic eruptions occur, violent explosions of steam. Usually, no one is hurt. But, in February 1979, two days after such an explosion, there was an earthquake in the Dieng region. Fearing another eruption, the inhabitants of the nearby village of Kepucukan fled, but they did not get far. During the earthquake a new fissure had broken open in the ground, out of which poured large amounts of carbon dioxide. The gas cloud caught up with the fleeing people, all of whom suffocated. Between 142 and 182 people were reported to have died in the catastrophe.

On New Zealand's North Island there are not only gas vents, mud pools and hot springs but also geysers – springs that eject boiling water and steam at regular intervals. Larger groups of these natural hot fountains are only found in four other areas on earth: in Yellowstone National Park in the United States; on the Kamchatka Peninsula in Siberia, in the far northeast of the Russian Federation; in Chile in South America; and on the volcanic island of Iceland in the North Atlantic. The word 'geyser' comes from the Icelandic, and means 'spring' or 'gush'.

Apart from high temperatures, geysers need narrow underground cavities filled with water. Like a gigantic boiler, the hot rock brings the water to boiling point. Because the temperature is higher in the deeper layers, the water there comes to a boil faster than in the upper layers. This is when pressure comes into play: due to the mass of cooler water in the upper cavities, the boiling water below cannot evaporate and instead becomes hotter and hotter. As soon as the water in the cooler upper section of the chamber also begins to boil and starts to evaporate, the pressure on the overheated water in the deeper section is relieved. Freed from this obstruction, the overheated water expands, suddenly turns into steam and shoots upwards through the narrow cavities. The geyser erupts.

The water of a hot spring collects in the dips and hollows of the thick sinter crust that covers the Whakarewarewa geyser field on the outskirts of Rotorua on New Zealand's North Island (right). Rain and meltwater fill the crater of the Ruapehu volcano, which lies around 140 kilometres south of Whakarewarewa in Tongariro National Park (far right). Fumaroles streaming out of the crater floor have turned it into an acid lake.

With every eruption, the underground pipes empty and the jet of hot water and steam dissipates. Cool groundwater flows in and refills the cavities. The process begins afresh.

The most powerful geyser ever known shot fountains of water and steam 460 metres into the air, and occurred as a result of a large eruption in 1886 at the Tarawera fissure, in the volcanic region of Taupo in New Zealand. The lava that flowed out dammed up a lake and blocked up a valley. From 1902 onwards, a geyser began to gush up every few hours in this valley, a short distance from the lake – this was Waimangu, the most powerful geyser of all time. However, when the water level of the lake began to sink in 1904, Waimangu stopped erupting, and has not been active since. The most spectacular active geysers in New Zealand today are in the Whakarewarewa geothermal area in the city of Rotorua. The most powerful, at 30 metres high, is Pohutu – the name means 'explosion' in Maori. Next to it is the Prince of Wales Feathers geyser, so named in 1900 during a visit by the then Prince, whose heraldic badge the geyser's fountain resembles. Both geysers erupt between 10 and 20 times a day.

Water can cause terrible catastrophes around volcanoes. If, for example, an acid lake overflows or heavy rainfall sweeps away fresh layers of ash or a glaciated volcano erupts and large amounts of ice rapidly melt, disastrous lahars or mudslides can result. Mount Ruapehu on New Zealand's North Island is notorious for its volcanic mud flows. Covered with an ice cap year round, it hides a hot acid lake in its crater. The worst eruption occurred here on Christmas Eve, 1953. Some of the ice melted due to the eruption and the crater wall burst. The water flowing out of the acid lake tore loose rocks and soil with it and became a mudslide that poured down over the flanks of the volcano for kilometres. It destroyed a railway bridge over which the night train to Wellington was travelling, and 151 people lost their lives. Since this disaster, Ruapehu has been very closely observed, particularly because the glacier is also used for skiing.

The heat given off by magma in the bedrock brings groundwater to the boil. It rises to the surface and emerges in the form of fumaroles, hot springs, geysers and mud pools (illustration, right). Fumaroles can hold a large amount of sulphur, which they deposit around their exit holes, as in the crater of the Mutnovsky volcano on Russia's Kamchatka Peninsula (far right).

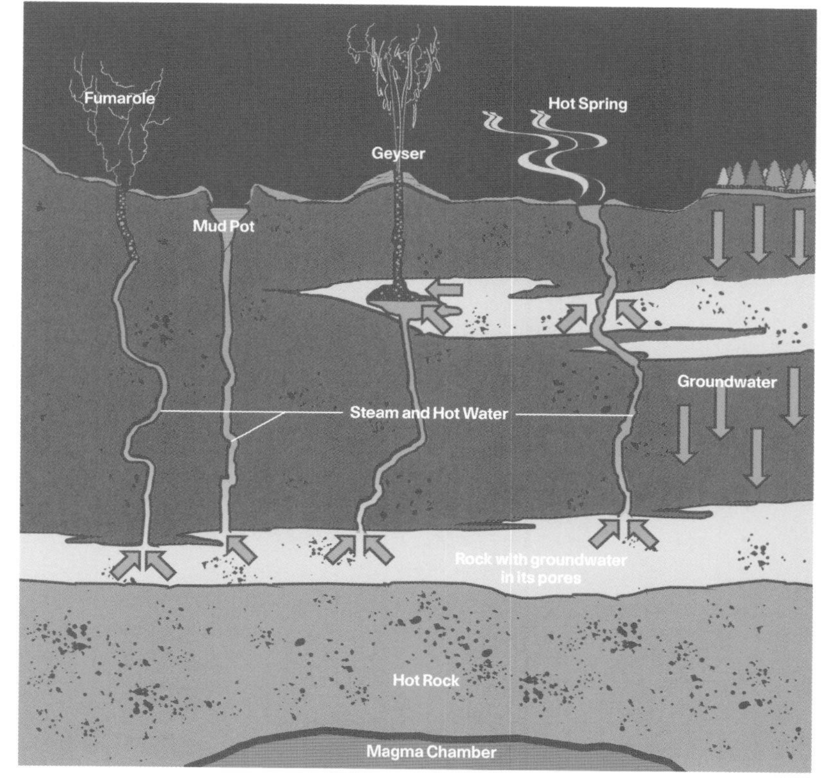

Atolls

Atolls – those circular coral islands with shallow lagoons in their centres – flummoxed the naturalists of the nineteenth century. There were heated debates as to how these islands in the middle of tropical oceans, far from the continents, could have developed from the depths of the tropical waters. Even the great English naturalist Charles Darwin (1809–82) addressed the issue, and as early as 1837, during his survey voyage on the *Beagle*, he developed a theory that is still held to be true today. Atolls, he believed, were built up by corals over the course of millions of years on volcanoes rising out of the water as islands. After the volcanoes became extinct, they gradually sank into the sea, while the coral reefs that encircled them continued to grow, so that finally just the ring of the reef with a lagoon in the middle remained.

His opponents were convinced that there must have been shallow sandbanks in the large tropical oceans on which corals built mighty circular reefs. As soon as the coral construction had reached a circumference that no longer allowed enough fresh water to penetrate the centre of the reef, the corals there gradually died off. The middle of the reef began to disintegrate and the lagoons that are so typical of atolls were formed.

It was not until around 1950 that Darwin's theory was proved correct by evidence that atolls do in fact sit upon extinct volcanoes. This evidence was a side effect of preliminary investigations in advance of hydrogen bomb tests in the South Seas. In order to analyse the bedrock of the Marshall Islands, the American government had boreholes drilled into the reef limestone of the Eniwetok Atoll. At a depth of more than 2,300 metres, the drills hit volcanic rock.

The 25 large atolls of the Maldives in the Indian Ocean, which run for a stretch of 860 kilometres due north of the equator, have a gigantic volcanic mountain range as their base. Seismic measurements taken by an oil company near the island of Bandos showed that volcanic rock at a depth of 2,100 metres is covered by reef limestone. The corresponding volcanoes had formed more than 100 million years ago, on the coat-tails, so to speak, of the then-isolated Indian Plate, which would only later fuse with the Australian Plate to create the Indoaustralian Plate. The Indian plate had just broken off from the supercontinent Gondwana and had begun to move north towards Eurasia. It passed a hot spot, a point where magma poured out from the depths of the earth. As the Indian Plate drifted over the hot spot, it was perforated by the rising magma, which functioned like a blowtorch, and volcanoes were created, one after the other.

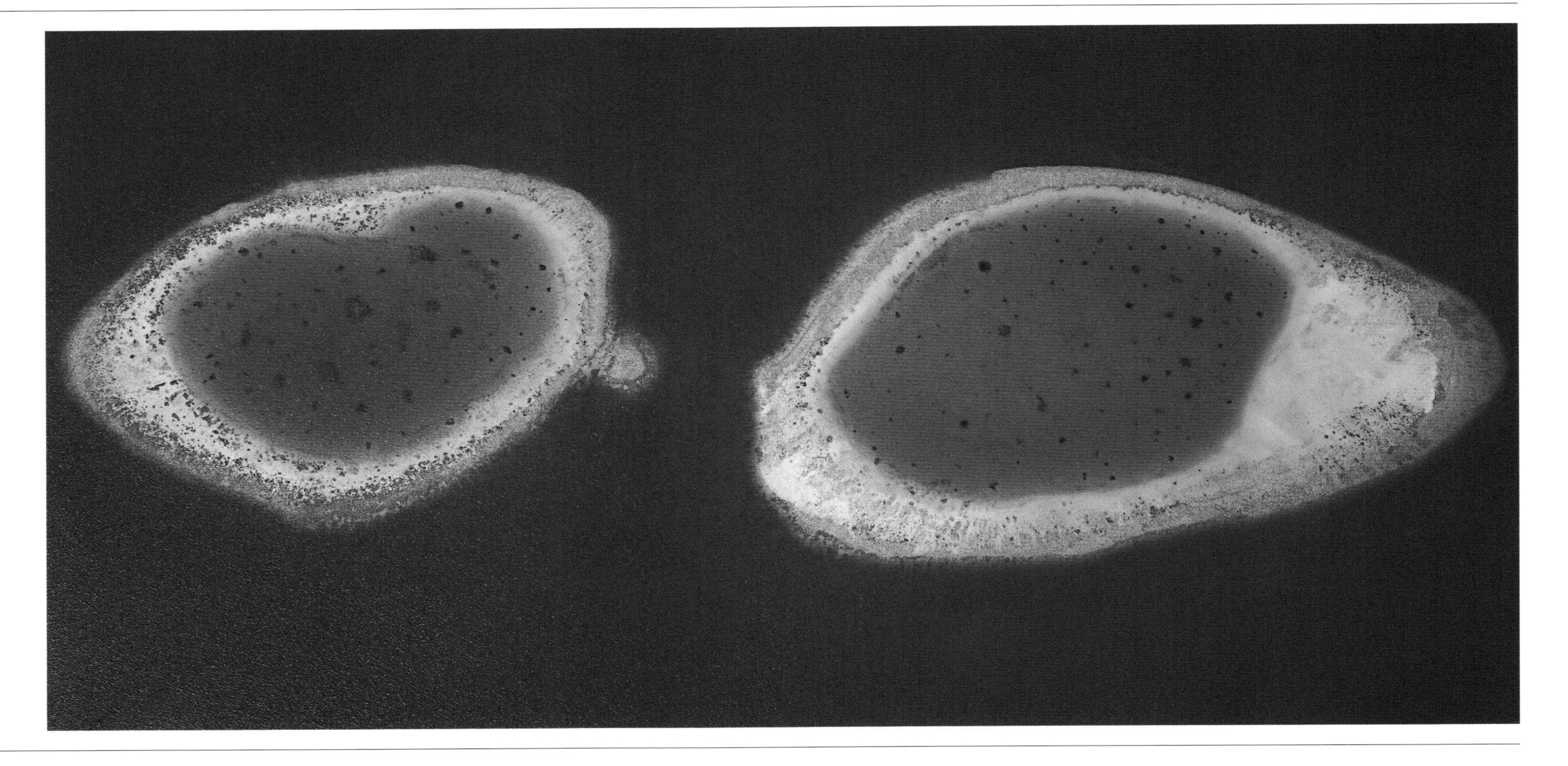

Both these lagoons encircled by coral reefs lie in the centre of the large Baa Atoll, about 120 kilometres from Male, the capital of the Maldives. They sit on a plateau of reef limestone more than 1,000 metres thick, which covers a sunken volcanic mountain range.

A similar process is currently playing out in the Pacific, resulting in the chain of islands that make up Hawaii. As soon as a volcano lost its connection with the hot spot, because it was being transported northwards on the drifting Indian Plate, it became extinct, sank slowly into the lithosphere and was finally covered in coral reefs.

The present-day appearance of the Maldives – a nation of more than 1,100 islands divided over 19 clusters of atolls – was, in addition, definitively determined during the ice ages. Over the last two million years, sea levels have varied by up to 100 metres, because a large volume of water has been held on the landmasses of the earth in the form of glaciers. The reefs have thus been dried out a number of times. The corals died off, the reefs weathered away. Large platforms of limestone were created, which were once again flooded with sea-water during the warm phases between ice ages. Primarily in the fresh, oxygen-rich sea-water around these platforms, new coral began to settle. A fringing reef developed, broken in places by the waves. Fresh sea-water now streamed into the lagoons, so that small coral islands could also begin to grow there.

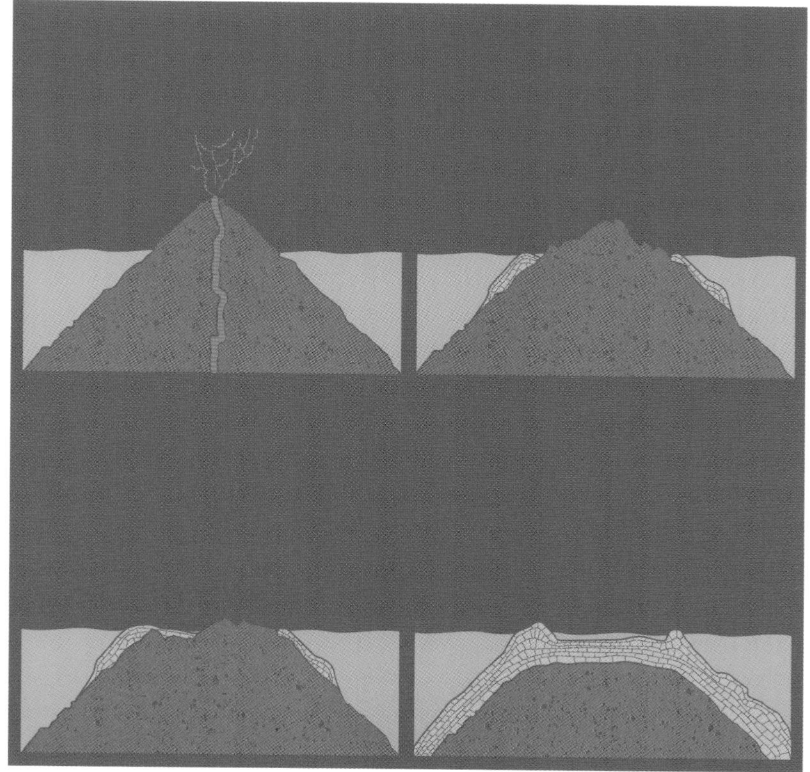

This schematic illustration shows the creation of an atoll (above right). When an island volcano in a tropical ocean becomes extinct and no more hot ash or lava flows stream over its flanks into the sea, coral begins to colonize its edges. As the volcano slowly sinks, the coral keeps growing towards the light near the surface of the water, creating a reef ring. The outer ring of the South Male Atoll, one of the great atolls of the Maldives, is only discernible from a great height (above left). Clear, fresh water enters the centre of the large atoll through channels, creating the perfect environment for coral reefs to be formed.

Glossary

Aa lava
Hawaiian term for a type of basaltic lava with a rough surface.

Acid lake
See **Crater lake**

Airfall
Volcanic ash which falls from an ash cloud or eruption column, also called ashfall.

Antimony
Scarce, silver-white chemical element, with the formula Sb, found in the earth's crust. In its natural form, it is usually combined with other elements (compounds). With **sulphur** it forms antimonite, and with oxygen, valentinite.

Ash
Fine pieces of rock and lava ejected explosively during volcanic eruptions.

Ash cloud
A cloud of ash ejected from a volcanic crater during an eruption, also called an eruption cloud.

Asthenosphere
A layer in the earth's mantle which is semi-soft, like modelling clay. It lies below the hard, rigid **lithosphere**, at a depth of more than 100 kilometres. In some regions it extends as deep as 700 kilometres. (See also **Interior of the earth**)

Atoll
A ring of coral reefs surrounding a central lagoon.

Bar
A unit indicating the physical force of atmospheric pressure.

Basalt
Dark-coloured, fine-grained volcanic rock, created when liquid lava rises from the depths of the earth. As it cools down, the molten rock sometimes splits into a series of separate polygonal surfaces, and solidifies into five- or six-sided basalt columns.

Black volcanic sand
Like all other rocks, lava is reduced to fine sand by weathering. This process is supplemented by the wave motion of the sea and the scraping of ice or glaciers, which not only crush the rock, but also carry it away and deposit it elsewhere.

Bomb
Pyroclasts or rocks ejected during a volcanic eruption with an average diameter greater than 64 millimetres.

Caldera
A roughly circular, steep-walled depression, often several kilometres in diameter, formed when the top of a volcano collapses into its empty magma chamber.

Carbonate
In purely chemical terms, carbonates are the salts of carbonic acid. Some kinds of carbonate are among the commonest rock-forming substances on earth. **Limestone**, for example, consists of calcium carbonate (chemical formula $CaCO_3$). It is chiefly deposited in sea-water. If CO_3 combines both with calcium and magnesium, it produces dolomite.

Carbon dioxide
Chemical compound which consists of two oxygen atoms bonded to a single carbon atom (chemical formula CO2). Found in the earth's atmosphere, it can also be expelled from the earth in eruption clouds and **mofettes**.

Chromium
One of the heavy metals (chemical formula Cr), this element is only present in very small quantities in the earth's crust. In ore deposits it occurs in the form of chromite, where chrome combines with iron and oxygen, and is brown in colour, whereas chromium alone is blue.

Cinder cone
A small volcano built primarily by ejected ash and **lapilli**.

Conduit
Passage through which magma flows in the volcano's interior.

Convection currents
Process within the mantle caused by heat transfer from the earth's core: hot semi-soft material rises and cooler semi-soft material sinks. This movement is most likely responsible for the motion of the earth's tectonic plates.

Coral reef
Structures of calcium **carbonate** built by coral animals on coasts between 30 degrees north and south of the equator. In order to develop, they need very clean, shallow sea-water, a great deal of light, and a water temperature between 20 and 30 degrees Celsius. Coral reefs can live up to a depth of 70 metres.

Core of the earth
See **Interior of the earth**

Crater
Round depression created through impact on or eruption of the earth's surface.

Crater lake
Lake formed when water collects in the crater of a volcano. When corrosive volcanic gases escape under water, it becomes an acid lake.

Crust of the earth
See **Interior of the earth**

Eruption
The volcanic release of lava and gas from the earth's interior onto its surface and into the atmosphere.

Eruption column
A mixture of volcanic ash, other pyroclastic material and hot gases which rises upward as a column above the erupting volcano. It can extend thousands of metres into the atmosphere.

Fault
A fracture in the earth's crust along which movement and displacement has occurred.

Fissure volcano
A crack in the earth's crust from which lava erupts.

Flue
Channel or system of tubes reaching many kilometres underground, through which glowing **magma** spills from the earth's interior to the surface, and is expelled through a **crater** (see **Volcanic vent**). After a phase of eruption the flue fills with stones and is gradually blocked by cooling lava – until it is blown open again with the next eruption.

Fumarole
Volatiles escaping from vents in the surface of lava flows and around the **craters** of **volcanoes** through which generally sulphurous, pungent-smelling gases and steam pours out. The gases can reach a temperature of 1,000 degrees Celsius. Because of the corrosive substances they contain, close proximity to them can irritate the mucous membrane (see **Precipitate**).

Geothermal spring
Spring from which warm water, often high in mineral content, emerges out of the depths of the earth. As soon as it cools down, the minerals are **precipitated** and deposited in a **siliceous** (or calcareous) **sinter** crust around the spring basin. If the water is hotter than human body temperature (37 degrees Celsius) it is called a 'hot spring'; if it ejects boiling water and steam it is termed a **geyser**.

Geyser
Natural **geothermal spring** that intermittently shoots fountains of steam and hot water, often several metres high, into the sky at intervals of minutes or hours. This phenomenon occurs only in volcanic zones, because only there is the temperature in the ground high enough to heat ground-water to boiling point time and again.

Geyserite
The name given to the **siliceous sinter** crusts that form around hot springs in geothermal areas (see **Geothermal spring** and **Precipitate**). When hot water flows downhill, terraces are formed by the sinter.

Glacier
Ice mass that forms on land surfaces in cold climatic regions, where temperatures are so low for most of the year that snow is able to collect regularly. Over time the layers of snow turn into ice under their own weight. Because pressurized ice behaves plastically, glacier tongues in the mountains creep down towards the valleys.

Global Volcanism Program
Program of the Smithsonian Institution, the renowned American educational and research organization, which seeks better understanding of the world's active volcanoes by documenting eruptions during the past 10,000 years.

Gneiss
One of the metamorphic rocks, most of which are very old, their original structure and composition transformed by extremely high pressure and temperatures within the earth. Gneiss is distinguished by its coarse-grained, layered texture, with different minerals arranged in parallel lines. Its main components are feldspar, quartz and mica.

Gondwana
Southern supercontinent formed after the supercontinent Pangaea broke up during the Jurassic period. It included what are now the South American, the African, the Indoaustralian and the Antarctic Plates.

Graben
Steep-walled trench structure between two parallel faults, also called rift.

Groundwater
Water located in pores and fractures of the rock beneath the earth's surface.

Guyot
Undersea mound, thought to be a volcanic cone truncated by water pressure during underwater eruptions into a circular, flat-topped shape. Many guyots lie in clusters on the floor of the Pacific, at depths of more than 200 metres. In the Afar Triangle in East Africa, which was once covered by a lateral branch of the Red Sea, guyots occur on the earth's surface.

Holocene
The most recent epoch in geological time, which begins after the ice age nearly 11,700 years ago and continues to the present.

Hot spot
A fixed centre of volcanic activity not related to plate tectonics. There are 40 to 50 hot spots on earth. The most active lie under Hawaii, Yellowstone and Iceland. The concentration of heat in the **lithosphere** and at the earth's surface is caused by a **mantle plume**.

Hot spring
See **Geothermal spring**

Island arc
Chain of volcanic islands along a **subduction zone** where oceanic **lithosphere** pushes beneath oceanic lithosphere.

Interior of the earth
The earth's interior is divided into three main regions. In the centre is the solid inner core, surrounded by the molten outer core. It has a radius of nearly 3,400 kilometres. The earth's mantle is a partly hard rock and partly semi-soft layer between the core and the earth's crust. The crust is the outer hard shell of our planet. There are two types – the 25- to 60-kilometre thick continental crust which forms the major landmasses, and the five-to seven-kilometre thick oceanic crust which forms the sea-floor (see also **lithosphere** and **asthenosphere**).

Iron oxide
Chemical compound of iron (Fe) and oxygen (O). The commonest iron oxides include the mineral hematite (Fe_2O_3). During the process of weathering, minerals containing iron, such as magnetite, produce hydrous iron oxides, with a range of colours from yellowish to rust-red.

Lahar
Hot mud and debris flow produced by volcanic eruptions. Lahars form predominantly on glaciated **volcanoes**, when a large quantity of volcanic ash (see **Lava**) is expelled. The hot ash that falls from the eruption cloud and settles on the **glacier** immediately causes the ice to melt. Then, laden with ash and stones, the mud flow rushes down into the valley.

Lapilli
Pea- to walnut-sized lava fragments ejected during a volcanic eruption.

Lava
As soon as **magma**, hot molten rock within the earth's interior, reaches the surface during the eruption of a **volcano**, it is termed 'lava'. It can flow out into streams or – in explosive eruptions – blow apart into fine particles, which are described as volcanic ash and can rise several kilometres into the air as an ash-cloud. Glowing lava generally has a temperature between 1,000 and 1,200 degrees Celsius. The rock formed when it cools down and solidifies is also called lava.

Lava dome
Dome-shaped mound of viscous lava slowly squeezed from the vent of a **stratovolcano**.

Lava tube
Tunnel in hardened lava that acts as a horizontal conduit for lava flowing from a volcanic vent. Lava tubes allow fluid lavas to advance great distances.

Limestone
Umbrella term for rocks consisting of **carbonates**, most of them made up of calcium and magnesium carbonate. They often contain fossils of prehistoric plants and animals, as well as shell fragments.

Lithosphere
Rigid outer part of the earth, including the crust and the uppermost part of the mantle. The plates drifting on the semi-soft **asthenosphere** are made of lithosphere (see **Interior of the earth**).

Maar
Shallow, flat-bottomed crater that forms above a volcanic vent when magma meets groundwater and steam explodes, mostly in a single event. Maars often fill with water and form lakes (see also **Phreatic eruption**).

Magma
Hot molten rock, usually of silicate composition, within the earth, which can rise to the surface through volcanic **flues**. As soon as it flows out of a **volcano**, the molten rock is called **lava**. But the magma can also become blocked on its way to the surface of the earth, and slowly harden. This leads to the formation of various so-called intrusions.

Manganese
Element with the chemical formula Mn. It is, after iron, the second commonest heavy metal in the earth's crust. Combined with oxygen, it forms the group of minerals called the manganese oxides. In deserts, manganese oxides, along with **iron oxides**, sometimes form a thin crust on the underlying rock surface known as 'desert varnish'.

Mantle of the earth
see **Interior of the earth**

Mantle plume
An upstream of hotter than normal magma that ascends towards the surface and pushes against the solid crust, where it may lead to volcanic activity. The plumes may originate as deep as the mantle-core boundary and are observed in the **lithosphere** and at the surface as **hot spots.**

Mercury
Element with the chemical formula Hg, also known as quicksilver. It is the only metal on earth that is liquid at room temperature. In nature, quicksilver is chiefly found in combination with **sulphur** in the mineral cinnabar, but it is also present in **volcanic gases**.

Mid-ocean ridge
Mountain chains on the sea-floor of all major oceans, thousands of kilometres long and some hundred kilometres wide, mostly with **grabens** at their axis. They are formed by upwelling magma and run along spreading zones.

Mofette
Cold to warm volcanic gas consisting mainly of **carbon dioxide** which leaks from cracks in the crust in volcanic areas. Mofettes often announce the final phase of volcanic activity. They can be dangerous, as carbon dioxide is heavier than oxygen and pushes it aside, causing suffocation.

Mud flow
See **Lahar**

Mud pool
Depressions in geothermal areas filled with boiling mud.

Obsidian
Rapidly cooled and solidified volcanic rock, which can look like black glass.

Pahoehoe lava
Hawaiian term for a type of lava with a smooth, rope-like surface.

pH value
Measure of the acidity or alkalinity of solutions, dependent on the concentration of hydronium ions. The higher the content of hydrogen ions, the lower the pH value and the more acidic the liquid. A solution with a pH value of 1 is considered highly acidic, one with a pH value of 7 is neutral and one with a value of 11–14 is highly alkaline.

Phreatic eruption
Explosion of steam, water, ash, rock and volcanic **bombs** that occurs when rising magma meets groundwater in the earth's crust.

Plate tectonics
Now firmly-established theory that the **lithosphere** is broken into chunks or plates which float on the semi-soft, hot **asthenosphere**. They change position and size over time.

Precipitate
A solid forced out of a liquid or from vapour as the result of a chemical reaction or a change in temperature or pressure. For example, when the water from **geothermal springs**, high in mineral content, emerges at the surface of the earth and cools, the minerals separate from it and form crusts (see **Siliceous sinter**, **Fumarole** and **Geyserite**).

Pseudocrater
Small craters formed by steam explosions when a lava-stream crosses swampy regions.

Pumice
A pale glassy volcanic rock with many cavities and pores, made by gases escaping through viscous lava. It is so light that it floats in water.

Pyroclastic flow
A hot, hundreds of degrees Celsius, mixture of volcanic gas, **ash** and rock fragments that travels at great speed down volcanic slopes. It forms when an **eruption column** collapses or when a **lava dome** collapses or explodes.

Glossary

Pyroclastic material
Volcanic rock that is ejected into the air during an eruption. It includes **ash**, **lapilli**, **bombs** and blocks.

Rift valley
A long trough bounded by parallel faults, formed when the earth's crust is stretched apart and a central section drops downwards.

Rift zone
A region of the **lithosphere** where extension leads to faulting. **Rift valleys** are the result of this process on continents. Sea-floor spreading occurs along active submarine rift zones.

Ring of Fire
Area around the Pacific ocean, around 40,000 kilometres long, where large numbers of volcanic eruptions and earthquakes occur. It is the result of the movement and collision of some **lithospheric** plates.

Scoria
Bubbly, porous fragments of **lava** expelled from **craters** during explosive volcanic eruptions. Hot scoria settles on the flanks of the volcanic cone, where it bakes into solid form.

Shield volcano
A broad, gently sloping volcano built from fluid lava streams. The biggest volcanoes on earth are shield volcanoes.

Silica
Minerals consisting of silicon and oxygen. They occur in at least nine different forms. The commonest of these is quartz.

Siliceous sinter
White, porous **silica** incrustation **precipitated** by **geothermal springs** and **geysers**.

Solfatara
Hot sulphurous gases escaping from cracks in the ground of geothermal areas. The name comes from the Latin *sulpha terra* which means land of sulphur. Solfatara is also the name of a wide shallow crater with many such gas vents near Naples, southern Italy.

Spreading zone
Divergent plate boundary. See also **Mid-ocean ridge** and **Rift zone**.

Stratovolcano
Steep-sided, cone-shaped volcano composed of **pyroclastic material**, **ash** layers and viscous lava flows that occurs at **subduction zones**.

Strombolian eruption
Low-level volcanic activity, named after the Stromboli volcano in the Thyrrenian Sea, Italy. Characterized by fountains of incandescent **ash**, **lapilli** and lava **bombs** which are thrown 10 to hundreds of metres high.

Subduction zone
A long narrow zone where one **lithosphere** plate decends beneath another and melts at a depth of about 100 kilometres. The magma which rises and erupts creates a chain of **stratovolcanoes**.

Sulphur gas
Gas, often foul-smelling and sometimes yellow in colour, that flows from the **craters** and fissures of **volcanoes**. These demonstrate that the volcano, even if it is not currently expelling **lava**, is still active (see **Fumarole**).

Supervolcano
Volcano with a huge magma chamber which can emit more than 1,000 cubic kilometres of material within one eruption – 10 times more material than an average volcanic eruption. The **pyroclastic material** and gases ejected into the atmosphere could probably cause climate change and mass extinctions.

Tectonic plates
See **Plate tectonics**

Tectonics
All processes that cause deformations in the earth's crust and allow new structures to form. They include both small deformations such as folds or fissures in rock, and large-scale changes like continental drift or the formation of mountain ranges, sea basins or large trough faults, of which the Rift Valley in East Africa is an example.

Tsunami
Japanese word for a long-wave-length ocean wave produced by an earthquake, landslide or volcanic blast. It reaches its greatest height in shallow waters before crashing on land. Often incorrectly called a tidal wave.

Tuff
Volcanic layers built from remnants of **pyroclastic flows**.

Volcanic gases
Volcanoes emit **fumaroles** (sulphur gases) and dangerous **carbon dioxide**, a colourless, odourless gas that is so heavy it drives the oxygen out of the layers of air closest to the ground, and can lead to suffocation.

Volcanic vent
Opening in a **volcano** from which glowing streams of **lava**, ashclouds or gases have emerged, or are still actively emerging (see **Flue**).

Volcano
A vent in the earth's crust through which **magma** and gases emerge. Volcanoes with a central vent are usually conical in shape. Those with several centres of eruption or an eruptive fissure form an elongated massif.

Index

Index

Phaidon Press Limited
Regent's Wharf
All Saints Street
London N1 9PA

Phaidon Press Inc.
180 Varick Street
New York, NY 10014

www.phaidon.com

First published 2009

ISBN 9 780 7148 5700 8

A CIP catalogue record for this book
is available from the British Library.

Translated by Ben Fergusson
Illustrations by Crush
Designed by Untitled
Printed in China

Authors' Acknowledgements
We would like to express our thanks
to the volcanologists who supported this
book with their knowledge of the earth's
fiery mountains; the pilots who circled
around volcanic cones time and again,
flew close to hot spots with great skill,
and always brought us safely back to
the cool earth; the rangers who provided
assistance in national parks and nature
reserves; the expedition organizers who
stood helpfully by us in extreme
volcanic regions; the many local people
who helped us secure flying permits
and overcome administrative hurdles;
Thomas Regnat for contributing to the
preparation of the picture files; and to
Alex Stetter, Denise Wolff, Paul McGuinness
and Amanda Renshaw at Phaidon, whose
great enthusiasm for the subject matter
made this book possible.